T0214002

# SpringerBriefs in Applied Sciences and Technology

SpringerBriefs present concise summaries of cutting-edge research and practical applications across a wide spectrum of fields. Featuring compact volumes of 50 to 125 pages, the series covers a range of content from professional to academic.

Typical publications can be:

- A timely report of state-of-the art methods
- An introduction to or a manual for the application of mathematical or computer techniques
- A bridge between new research results, as published in journal articles
- A snapshot of a hot or emerging topic
- An in-depth case study
- A presentation of core concepts that students must understand in order to make independent contributions

SpringerBriefs are characterized by fast, global electronic dissemination, standard publishing contracts, standardized manuscript preparation and formatting guidelines, and expedited production schedules.

On the one hand, **SpringerBriefs in Applied Sciences and Technology** are devoted to the publication of fundamentals and applications within the different classical engineering disciplines as well as in interdisciplinary fields that recently emerged between these areas. On the other hand, as the boundary separating fundamental research and applied technology is more and more dissolving, this series is particularly open to trans-disciplinary topics between fundamental science and engineering.

Indexed by EI-Compendex, SCOPUS and Springerlink.

More information about this series at https://www.springer.com/series/8884

Bupe G. Mwanza · Charles Mbohwa

# Sustainable Technologies and Drivers for Managing Plastic Solid Waste in Developing Economies

Springer

Bupe G. Mwanza
Graduate School of Business
University of Zambia
Lusaka, Zambia

Charles Mbohwa
Department of Operations Management
University of Johannesburg
Johannesburg, South Africa

ISSN 2191-530X              ISSN 2191-5318  (electronic)
SpringerBriefs in Applied Sciences and Technology
ISBN 978-3-030-88643-1         ISBN 978-3-030-88644-8  (eBook)
https://doi.org/10.1007/978-3-030-88644-8

This Springer imprint is published by the registered company Springer Nature Switzerland AG
The registered company address is: Gewerbestrasse 11, 6330 Cham, Switzerland

*This book is dedicated to manufacturers and solid waste researchers with the desire to contribute to the circular economy.*

# Preface

This book examines research issues and trends on Sustainable Waste Management (SWM) technologies used for managing End-of-Life (EoL) post-consumer Plastic SOLID Wastes (PSWs). It provides the way forward to resource management of EoL post-consumer plastic wastes in developing economies by making reference to research in developed economies. It is directed toward individuals who are researchers and innovators in the Solid WASTE Management (SWM) space or have a significant research leadership role in it. The book appeals to researchers in environmental management and contributes to research in SWM. Research-based and informed sustainable options for managing PSWs are presented in a managerial, scientific and engineering approach to enable decision makers in industry and the economy to implement appropriate strategies toward achieving sustainable resource management. The book further integrates Waste Management (WM) drivers and the technologies in order to establish a sustainable relationship for managing PSWs from an environmental, economic and social perspective. It interrogates the use of sustainable technologies for managing post-consumer PSW in developing economies while highlighting the drivers that would influence the key stakeholders to achieve sustainability in the plastic industry.

Chapter 1 provides a critical review of SSWM from the developed economy perspective while making an inference to developing economies. Critical definitions of SW are provided with a view of making decision makers adopt wastes in resource utilization programs.

Chapter 2 focuses on recent trends in the field of PSWs management. It provides an overview of post-consumer PSWs while highlighting their sources. The merits and challenges of managing PSWs in developing economies are presented.

Chapter 3 presents a review of technologies for managing PSWs. Recycling, energy recovery, composting and landfilling are reviewed from an economic, environmental and social perspective. Recommendations on how waste managers, engineers and other stakeholders can improve PSWs management are provided from a technological perspective. The challenges of implementing these technologies are discussed for the purpose of providing insights to those involved PSWs management.

Chapter 4 reviews in detail the aspect of sustainability by focusing on its three pillars. Further, sustainability is evaluated by addressing the indicators in the SWM arena. To expand the aspect of sustainability, a Life Cycle Assessment (LCA) of PSWs is integrated into the chapter in order to provide guidance to policy makers when considering technologies for implementation and legislation.

Chapter 5 describes in detail the drivers that affect the sustainable management of PSWs. These drivers are discussed by focusing on success stories in developed economies. The chapter demonstrates how these can work in developing economies.

Chapter 6 focuses on the drivers that should be integrated and a model is proposed. The drivers are suggested for developing economies in the management of PSWs.

Chapter 7 provides the precise contributions to SSWM.

Riverside (Kitwe), South Africa                                                    Bupe G. Mwanza
Johannesburg, South Africa                                                          Charles Mbohwa

# Contents

# Abbreviations

| | |
|---|---|
| BFBs | Bubbling fluidized beds |
| $CO_2$ | Carbon dioxide |
| CSR | Corporate social responsibility |
| ELFM | Enhanced landfill mining |
| EoL | End-of-life |
| EoU | End-of-use |
| EPA | Environmental Protection Agency |
| EPR | Extended producer responsibility |
| GHG | Greenhouse gas |
| IWCs | Informal waste collectors |
| IWS | Informal waste sector |
| LCA | Life cycle assessment |
| MRFs | Material recovery facilities |
| MSW | Municipal solid waste |
| NO | Nitroxide |
| OECD | Organization Economic Cooperation and Development |
| PAHs | Polycyclic aromatic hydrocarbons |
| PCDFs | Polychlorinated dibenzofurans |
| PE | Polyethylene |
| PET | Polyethylene terephthalate |
| PP | Polypropylene |
| PPW | Package and package waste |
| PS | Polystyrene |
| PSWs | Plastic solid wastes |
| PVC | Polyvinyl chloride |
| RCRA | Resource Conservation and Recovery Act |
| RL | Reverse logistics |
| SDG | Sustainable development goals |
| SSWM | Sustainable solid waste management |
| SW | Solid Waste |
| SWM | Solid waste management |

| UNEP | United Nations Environment Programme |
| VOCs | Volatile organic compounds |
| WM | Waste management |
| WMS | Waste management system |
| WTO | World Trade Organization |

# Chapter 1
# Sustainable Solid Waste Management: A Critical Review

The chapter provides an overview of sustainable solid waste management (SSWM). In order to understand SSWM, the definitions of solid wastes (SWs) are reviewed and gaps are identified. From the identified gaps, a new definition of plastic solid wastes (PSWs) is developed. Sustainable ways of managing PSWs are discussed and the 3Rs initiatives are considered. A number of research works from developed and developing economy perspectives are reviewed in order to present a clear understanding of the current situation with regards to PSWs management. An SSWM model adapted from a study on SW management is discussed in order to present the current trends of managing PSWs. The chapter provides simplified insights to the wastes managers, engineers and stakeholders engaged in SW management.

## 1.1 General

Population growth, rising standards of living and increasing urbanization because of technological innovations contribute to an increase in the amount and variety of solid waste (SW) generated. These changes can be attributed to the continuous pursuit of high-quality lifestyles, which resulted in increased consumption of resources. In reality, the generation of SW is a natural consequence of human life.

Increased consumption patterns result in increased production of SW and this results in related pollution and health challenges for nations. Globally, the estimated total annual SW generation is 17 billion tons, and by 2050, 27 billion tons will be generated [19]. These wastes have a direct effect on the social, environmental and economic aspects of nations. In this regard, sustainable solutions that consider techno-economic, socio and environmental aspects should be established. Solutions to SW-related challenges are managed using the waste management hierarchy by

B. G. Mwanza and C. Mbohwa, *Sustainable Technologies and Drivers for Managing Plastic Solid Waste in Developing Economies*, SpringerBriefs in Applied Sciences and Technology, https://doi.org/10.1007/978-3-030-88644-8_1

identifying the appropriate processes for use in order to achieve sustainability, and the EU Framework of [9] shows that the waste hierarchy is supported by the sustainability ranking. Consequently, solid waste management (SWM) initiatives are moving from waste disposal concepts to resource utilization. Couth and Trios [6] indicate that it is more sustainable to prepare SW for reuse and recycle processes compared to energy recovery and landfilling. The purpose of this book is to demonstrate and discuss sustainable technologies for managing plastic solid wastes (PSWs) in developing economies.

Management of SW in developing economies is achieved with numerous challenges and with the notion of sustainability, considerable efforts need to be established. With regard to resource utilization, the goal of any sustainable solution is achieving efficiency and maximization in resource utilization at every step of the process inclusive of the production and disposal stages. Further, challenges encountered in the waste supply chain should be addressed in totality because of the close dependence of stages on each other. This book addresses the aspect of achieving totality in managing PSWs by focusing on sustainable technologies in the context of developing economies. The technologies discussed complement each other by focusing on resource utilization.

Plastics are an integral aspect of our lives because of several properties that they possess. Several million tons of plastic products are manufactured annually and are mostly utilized for consumer packaging [3, 23]. Approximately 4% of the annual petroleum is utilized in the manufacture of plastics [5]. Further, 50% of plastics are manufactured into single-use packaging products [17]. This implies huge quantities of PSWs are generated on a yearly basis. Coupled with the SW-related challenges facing developing economies, PSWs are a major concern, especially with non-biodegradable properties. Identification of the proper methods for managing PSWs in the developing economies context is relevant. This book discusses sustainable technologies that benefit waste management experts, manufacturers and engineers. These technologies are also relevant to individuals pursuing projects in PSWs management for resource utilization purposes.

According to [10], 7.3% of plastic production is from the Middle East and Africa. In comparison with other economies, 7.3% of plastic production is on the lower side. Nevertheless, in developing economies such as Africa, less than 45% of SW is collected [16]. Further, 8–12% of municipal solid wastes (MSWs) are PSWs [16]. In addition, the composition of PSWs in MSWs is expected to increase to 13% by 2025. The proportional increase in PSWs generation in most developing economies has not affected proportional sustainable waste management solutions. In this regard, the book provides sustainable solutions to the number of challenges affecting developing economies.

As the global production of plastic products continues to increase, a direct and proportional increase in the amount of PSWs generated occurs. Therefore, it is important to develop sustainable solutions for managing PSWs from the environment bearing in mind the future generations. The book discusses the technologies and drivers for managing PSWs in developing economies. Practically a number of existing technologies for managing PSWs provide success stories in developed economies. Nevertheless, a direct application of the existing technologies in developing economies faced with different PSWs dilemmas may not provide a solution. The book discusses the technologies and drivers that work in managing PSWs in the developing economies context. Further, reference is made to successful PSWs management technologies from developed economies.

## 1.2 What is Solid Waste?

SW management is a critical element of urban management because of its linkage to human health and infrastructure development. The primary purpose of SWM strategies is to solve environmental, social and economic concerns related to improper management of waste [15, 33]. Historically, WM has existed to solve health and safety issues caused by wastes [26]. To achieve the primary goals of SWM, it is required that the stakeholders responsible for managing SW understand the meaning of waste. A number of definitions exist.

According to the German Waste Act (1972), waste is defined as *"portable objects abandoned by the generators"* [2].

The USA Resource Conservation and Recovery Act (RCRA, 1976) defines waste "in the form of liquid, gas or solid generated from any sector of the economy inclusive of community activities" [18].

In the New Zealand Waste Strategy, waste is defined as *"any material existing in any form of matter that is unvalued and discarded by the owner"* [7].

According to Ngoc and Schnitzler [22], *"waste consists of items generated and unwanted by the people or companies and require disposal."*

The definitions of waste from the above authors may have a common phrase of material discarded but the emphasis of each definition is different. The German Act makes mention of abandoned materials that require proper disposal for public protection. However, the USA definition focuses on the origin and type of waste. The New Zealand definition focuses on the types of waste. The definition by Ngoc and Schnitzler [22] highlights that wastes are all items not wanted by the generator or the owner. Clearly, from these definitions, it can be noted that waste has been labeled as materials or items generated by a process or people and is no longer needed but only fit for disposal. None of the definitions have considered waste as a material fit for adoption in another process as depicted by the waste management hierarchy (i.e. reuse, recycle and energy recovery).

In this book, *SW is defined as any discarded solid material or item unwanted by the owner or not fit for a process but can be reused or used as an input material in the manufacture of another useful product.*

Waste of any form should not be considered useless unless proven otherwise. For this reason, SWM around the World Cities [13] affirms solid waste is a resource, so recovery rates across the waste chains should be maximized. Therefore, the focus around SW in this book centers on maximizing the utilization of discarded materials or items for the purpose of achieving sustainability. This is achievable by understanding the technologies that are used in converting these SWs into useful resources.

SW is anything from unwanted bottles, papers or plastics. The different types of SW are found in markets, public institutions, hospitals, industries and many other waste streams. PSWs consist of different types of wastes and this book focuses on post-consumer packaging PSWs.

Post-consumer packaging PSWs consist of end-of-use (EoU) or end-of-life (EoL) plastic packages discarded by the owner or the user. With 50% of plastics being manufactured into packaging materials, a significant quantity of PSWs is generated. The challenge to the waste management experts is on how to sustainably manage these waste types. Understanding the appropriate technologies and strategies for managing PSWs is necessary to waste management experts. This book discusses the different types of PSWs and the existing technologies appropriate for management.

## 1.3  Sustainable Solid Waste Management

Sustainable solid waste management (SSWM) is an approach that addresses WM pressures through reusing, recycling and recovering resources as well as waste minimization. It utilizes environmentally and economically friendly strategies by putting waste separation and reduction practices at the center of the waste hierarchy [1, 29, 31]. Despite the existence of the WM hierarchy, SWM is still a challenge in a number of developing economies. The challenges include low budgets allocated for WM, increased waste generation, limited knowledge on WM at the different stages of the product life cycle and lack of sustainable WM systems [12].

As a result of these challenges, most developing countries turn to illegal disposals despite having restricted developed lands, reduced rates of waste processing and crumbling SWM infrastructure [25]. While these challenges in SW management continue, it is necessary that SSWM approaches are considered across a variety of country-specific scenarios. This implies that the implementation of SSWM approaches by the policy makers and waste management experts should be applied with consideration of SSWM practices. The SSWM practices should involve decisions that are operational, strategic and tactical across the management hierarchies. The practices of SSWM give consideration to the selection of suitable engineered landfills and waste treatment facilities [32], strategic capacity transformation for waste treatment and landfilling [14], appropriate collection schedules for each waste type and zone [4] and service zoning [21]. Application of SSWM practices to the management of SW results in positive results, however, waste managers and experts should align the selection of appropriate technologies to the SSWM approach and practices.

Understanding the concepts in SSWM is cardinal for waste experts and researchers because SWM is a visible, political and sensitive service having inadequacies within the service that results in numerous credibility implications in the public administration. Within the city government, SWM is a vital utility service on which public health and the beauty of the city depend. Proper management of SW is important for public health and sustainability reasons; hence requires serious attention from the relevant stakeholders.

Globally, the drive for sustainable efforts for reducing material consumption has shifted from SWM techniques of waste elimination from habitable areas [27] to resource utilization. SSWM embraces the 3R initiatives (reduce, reuse and recycle) as a way of achieving sustainability. Application of the 3R initiatives reduces the quantity of waste that enters the landfill [8]. In reality, SSWM is one aspect of achieving the notion of circular economy in which waste is no longer considered waste but a resource. Therefore, the technologies discussed in this book contribute to achieving the circular economy notion by applying SSWM initiatives. The 3Rs initiatives of SSWM are important for waste managers to understand each initiative that aims to contribute to sustainability in a different manner.

*Reduce initiative*: This initiative cuts across the product life cycle, from raw material generation, its manufacture, its utilization to the end-of-life. Waste reduction is placed at the top of the WM hierarchy, implying it is the most desirable initiative. In essence, all activities of an item have not been considered as waste earlier. This initiative basically means decreasing the quantity of waste produced by prolonging the product's life span through reuse; decreasing the amount of waste associated with those giving an environmental and public health impact; and decreasing the number of noxious substances in products [30].

*Reuse initiative*: This occupies second place on the waste management hierarchy. It basically focuses on the utilization of a product after its intended purpose has expired. This means it can be used for the same purpose as before or as an entirely new one. For example, the reutilization of beverage bottles or plastic bags (William 2005). The European Commission [30] indicates that reuse is an operation applied to a product or its component after the end of its initial life. In addition, using the product for the same purpose for which it was intended, including the continued use of a product returned at a waste collection point, manufacturer, distributor or recycler as well as reuse after refurbishment.

*Recycle initiative*: This option focuses on productive value addition in the stages of waste segregation, collection and processing [24]. Waste can be either produced into a similar product or a different one. It should be known that environmental considerations are key for recycling activities to be achieved. For example, energy consumption and pollution production of a recycling process should be lower than the utilization of virgin material. Other aspects such as marketability, cost and standing legislations should be ascertained.

Despite the fusion of the 3Rs initiatives in SSWM, waste experts, researchers and those charged with the responsibility of managing waste should understand that SSWM goes beyond technical aspects and includes sustainability elements as a way of ensuring the success of SW projects [35]. According to a study by [20], the key aspects of sustainability include legal, financial, administrative, socioeconomic, technical and environmental considerations. These aspects should be integrated to support the management of SW in the context of a particular region. Figure 1.1

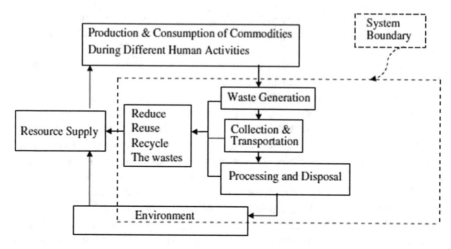

**Fig. 1.1**  Sustainable solid waste management system (adapted from [27])

depicts the schematic diagram of an SSWM system. According to this system, an environment is a place for resources and a drop for undesired materials.

According to Shedkar [27], a system for SSWM should have the capacity to meet societal needs such as the technological, financial and environmental, and assimilative capacity of any changes. Such a system should be pragmatic and provide realistic approaches for improvement. Without consideration given to whether the system will work in the context of that society, the development and adoption of that system will be in vain.

Waste experts should be aware that sustainable management of PSWs embraces the key elements and aspects of SSWM. The important principle is understanding how to utilize the 3R initiatives in managing PSWs. Figure 1.1 depicts a scenario applicable to all types of wastes, but what a waste manager in charge of PSWs needs to know is how to integrate the system elements (i.e. waste generation, collection, separation etc.) and the key aspects to sustainably manage PSWs. Waste managers should know that PSWs are recovered when diverted from landfills; otherwise, source separation should exist. The quantity of PSWs that enters the waste management system (WMS) can be minimized by decreasing the number of plastic materials in the first place and reducing the use of plastic materials in products (e.g. utilization of lighter packaging instead of heavy ones or packaging down-gauging). Designing plastic products to enable remanufacturing, repairing or reusing contributes to fewer EoL products in the waste stream. Once the plastic is captured in the waste stream, recycling is one of the processes that can be used to recover the materials so that a new product can be manufactured. Energy recovery has not been depicted in Fig. 1.1 as a material recovery process. It is a concept that can be included in the model when the calorific value of the material utilized is under controlled combustion (i.e. fuel). In addition, energy recovery has lesser environmental performance compared to recycling but demand for virgin material is not reduced. The concept of the 4Rs strategy in the WM arena focuses on environmental desirability in which reduce, reuse and recycle of materials is the key, while recovery is more on energy and land-filling is the least considered strategy. The possibility of the same polymer to cascade via multiple stages such as manufacturing to reusing via recycling and eventually energy recovery is high. In developing economies, challenges of managing PSWs still exist even though the 4Rs strategy forms part of an economic activity for certain sectors of societies. For this reason, most of the developing economies resort to open dumping as there is a realization that the 3R initiatives of SSWM are inadequate.

Application of SSWM concepts in developing economies is still in its infancy as several studies have indicated that SWM including management of PSWs encounters a number of challenges [11, 25, 28, 34]. The existence of these challenges does not remove the responsibility placed on waste experts to identify sustainable solutions. According to a study conducted on sustainable solutions for SWM [22], the study identifies the use of "zero waste" strategy to promote material recycling, reusing, prevention and redesign of the product's life cycle. This emerging trend in SWM encompasses the 3Rs initiatives of SSWM. For this solution to be applicable and workable in developing economies, it must be developed in the societal context of the countries.

# References

1. Authority, Environment Protection, *Environment Protection Authority Annual Report 1 July 2014 to 30 June 2015*. Annual Report, Adelaide SA 5001: European Protection Authority (2015)
2. B. Biltewski, G. Hardtle, K. Marek, *Waste Management* (Springer, Berlin Heidelberg New York, 1997)
3. I. Blanco, End-life prediction of commercial PLA used for food packaging through short term TGA experiments: real chance or low reliability, Chin. J. Polym. Sci. **32**, 681–689 (2014)
4. F. Chu, N. Labadi, C. Prins, A scatter search for the periodic capacitated arc routing problem. Eur. J. Oper. Res. **169**(2), 586–605 (2006)
5. Consumption, British Plastic Federation Industry 2008 Oil, Accessed May 20, 2016. http://www.bpf.co.uk/Oil-Consumption.aspx (2008)
6. R. Couth, C. Trois, Sustainable waste management in Africa through CDM projects. Waste Manage. **32**, 2115–2125 (2012)
7. Environment, Ministry for the 2002, *Ministry for the Environment New Zealand*. Accessed May 01, 2017. https://www.mfe.govt.nz/sites/default/files/waste-strategy-review-progress-mar07.pdf
8. Environment, Ministry of. 2006, *Ministry of Environment Government of Japan*. Accessed November 21, 2017. https://www.env.go.jp/en/wpaper/2006/fulltext.pdf
9. EU2008/98/EC, *European Commission*. Accessed November 20, 2018 (2008). https://ec.europa.eu/environment/waste/framework/
10. G. Gourmelon, *Global Plastic Production Rises, Recycling Lags* (WOrldWatch Institute, 2015). May 12. Accessed 12 Jan 2019
11. D. Grazhdani, Assessing the variables affecting on the rate of solid waste generation and recycling: an empirical analysis in Prespa Park. Waste Manage. **48**, 3–13 (2016)
12. L.A. Guerrero, G. Maas, W. Hogland, Solid waste management challenges for cities in developing countries. J. Waste Manag. **33**, 230–232 (2012)

13. HABITAT, UN, *Solid Waste Management Around the Cities* (2010). Accessed 2018. https://sswm.info/sites/default/files/reference_attachments/UN%20HABITAT%202010%20Solid%20Waste%20Management%20in%20the%20Worlds%20Cities.pdf
14. L. He, G.-H. Huang, G.-M. Zeng, H.-W. Lu, Identifying optimal regional solid waste management strategies through an inexact integer programming model containing infinite objectives and constraints. J. Waste Manage. **29**(1), 21–31 (2009)
15. K.R. Henry, Z. Yongsheng, D. Jun, Municipal solid waste management challenges in developing countries—Kenyan case study. Waste Manage. **26**, 92–100 (2006)
16. D. Hoornweg, P. Bhada-Tata, *What a Waste: A Global Review of Solid Waste Management. Urban Development Series Knowledge Papers* (World Bank, Washington, DC, 2012)
17. J. Hopewell, R. Dvorak, E. Kosior, Plastics recycling: challenges and opportunities. Phil. Trans. R. Soc. B **364**, 2115–2126 (2009)
18. Institute, Legal Information, Title 42—Chapter 82. August 15. Accessed May 02, 2018 (2003). http://www4.law.cornell.edu/uscode/42/ch82.html
19. T. Karak, R.M. Bhagat, P. Bhattacharyya, Municipal solid waste generation, composition, and management: the world scenario. Crit. Rev. Environ. Sci. Technol. **42**(15), 1509–1630 (2012)
20. A. Van de Klundert, Integrated Sustainable Waste Management: the selection of appropriate technologies and the design of sustainable systems is not (only) a technical issue, in *Inter-Regional Workshop on Technologies for Sustainable Waste Management*, held 13–15 July (CEDARE/IETC, Alexandria, Egypt, 1999)
21. M.C. Mourão, A.C. Nunes, C. Prins, Heuristic methods for the sectoring arc routing problem. Eur. J. Oper. Res. **196**(7), 856–868 (2009)
22. U.N. Ngoc, H. Schnitzer, Sustainable solutions for solid waste management in Southeast Asian countries. Waste Manage. **29**, 1982–1995 (2009)
23. S. Papong, P. Malakul, R. Trungkavashirakun, P. Wenunun, T. Chomin, M. Nithitanakul, Comparative assessment of the environmental profile of PLA and PET drinking water bottles from a life cycle perspective. J. Clean Prod. **65**, 539–550 (2014)
24. S. Pattnaik, M.V. Reddy, Assessment of municipal solid waste management in Puducherry (Pondicherry), India. Resour. Conserv. Recycl. **54**, 512–520 (2010)
25. A.S. Permana, S. Towolioe, A.A. Norsiah, S.H. Chin, Sustainable solid waste management practices and perceived cleanliness in a low income city. Habitat Int. **49**, 197–205 (2015)
26. C. Ponting, *A Green History of the World* (Sinclair-Stevenson, Great Britain, 1991)
27. V.A. Shekdar, Sustainable solid waste management: an integrated approach for Asian countries. Waste Manage. **29**, 1438–1448 (2009)
28. Tacoli, urbanization, gender and urban poverty: paid work and unpaid carework in the city, in *International Institute for Environment and Development: United Nations Population Fund* (United Nations, London, UK, 2012)
29. Tool, Sustainable Facilities, *System overview: Solid waste management hierarchy* (2014). Accessed November 18, 2018. https://sftool.gov/explore/green-building/section/57/solid-waste/system-overview
30. E. Union, *Being Wise with Waste: The EU's Approach to Waste Management* (Publications Office of the European Union, Luxembourg, 2010)
31. US-EPA, *Non-hazardous Solid Waste Management Hierarchy* (2013). Accessed 12 June 2019. http://www.epa.gov/solidwaste/nonhaz/municipal/hierarchy.htm
32. G. Wang, L. Qin, G. Li, L. Chen, Landfill site selection using spatial information technologies and AHP: a case study in Beijing, China. J. Environ. Manage. **90**(8), 2414–2421 (2009)
33. D. Wilson, A. Whiteman, A. Tormin, *Strategic Planning Guide for Municipal Solid Waste Management* (World Bank, Washington, DC, 2001)
34. T.B. Yousuf, M. Rahman, Monitoring quantity and characteristics of municipal solid waste in Dhaka City. Environ. Monit. Assess. **135**, 3–11 (2007)
35. C. Zurbrügg, M. Gfrerer, H. Ashadi, W. Brenner, D. Küper, Determinants of sustainability in solid waste management—the Gianyar waste recovery of sustainability in solid waste management-project in Indonesia. J. Waste Manage. **32**, 2126–2133 (2012)

# Chapter 2
# Post-consumer Plastic Solid Wastes: Recent Research Trends

An overview of the different types of plastics is presented in this chapter. Plastics are categorized into thermoplastics and thermosetting. The different types of plastics under the two categories and the purposes for which they are manufactured are discussed. For the purposes of understanding the existing post-consumer plastic products, an overview of post-consumer plastic products is presented. In order for managers, engineers and other stakeholders to understand the sustainable ways of managing PSWs, a comparison of plastic production from an income-level perspective of countries is discussed. Further, the chapter discusses the challenges developing economies are facing in managing PSWs. In order to present applicable solutions to managers, engineers and policy makers, current trends of managing PSWs are discussed.

## 2.1 General

Plastics are durable and inexpensive materials that are readily molded into numerous products with a range of applications. Plastics are manufactured from substances such as natural gas and petroleum. For this reason, they are highly polymerized compounds that consist of hydrogen and carbon. Further, crude gasoline is used as one of the raw materials in the manufacturing of plastics.

Based on the heating behavior, plastics are categorized into thermoplastics and thermosetting. Thermoplastics soften when heated as a result of the molecular motion. The process of continuous heating and cooling affects the ability of plastics to be molded into various shapes. Thermoplastics consist of polyethylene terephthalate (PET), low-density polyethylene (LDPE), polyvinylchloride (PVC), high-density polyethylene (HDPE), polypropylene (PP) and polystyrene (PS). Weak molecular

© The Author(s), under exclusive license to Springer Nature Switzerland AG, part of Springer Nature 2022
B. G. Mwanza and C. Mbohwa, *Sustainable Technologies and Drivers for Managing Plastic Solid Waste in Developing Economies*, SpringerBriefs in Applied Sciences and Technology, https://doi.org/10.1007/978-3-030-88644-8_2

motions are experienced in thermosetting plastics. Continuous heat treatment causes thermosetting plastics to undergo chemical reactions. The process of heating enables the formation of a molecular 3D matrix structure that cannot be softened by further heating.

Plastics possess quite a number of favorable properties compared to other materials. The properties include lightweight, durability, robustness, resistance to rust and many others. The properties possessed by plastics have made them occupy every spectrum of our lives. A study conducted by the British Federation [9] indicates that around 4% of the annual petroleum manufactured is changed into plastics. Further, 3–4% is used for the provision of energy for manufacturing. Plastics continue to dominate in their usage as post-consumer packaging materials despite the current consumption of non-renewable resources. Approximately 50% of plastics are used as once-off disposable products such as packaging or consumer items [15]. This implies that 50% of the manufactured plastics are disposed of as wastes after their intended purpose.

Numerous products previously manufactured from materials such as glass, paper, wood or metal are now manufactured from plastic materials. Al-Salem et al. [5] indicate that the favorable properties possessed by plastics have presented this opportunity. In developing and developed economies, the changes in the use of plastics have contributed to the economic growth of plastic manufacturing companies. For example, the annual growth rate for the polymer industry is expected to grow by 3.9% from 2015 to 2020 [40]. The design flexibility, lightweight, durability and low-cost properties of the plastics enable the increase in the manufacture of plastics into various products. According to Qualman [48], nearly 400 million tons of plastics are manufactured per year. This implies that per day more than a million kilograms are produced. The growth in the manufacture of plastic materials and products has contributed to the technical and economic development of the plastic industries. Figure 2.1 depicts the projected global plastic production by 2050.

Figure 2.1 shows that the tonnage of plastic production has doubled in less than two decades. Developing economies such as the Middle East and Africa are not an exception to the anticipated growth. Despite the advantages presented by plastics, the projected levels of production are clearly an indication of future challenges attributed to plastics. It is important to indicate that at the current production levels, plastics present a number of problems. A number of plastic-related problems are discussed in this chapter.

Table 2.1 depicts the global composition of wastes by income levels. The projected waste composition by 2025 indicates the growth in the composition of plastic wastes at all income levels. Several factors may be attributed to the increase in PSW composition. Qualman (2017) indicates that only 18% of plastics used in the automobile and construction industry are recycled globally. For packaging plastics, only 14% is recycled globally. Factors such as the exclusion of plastics at the collection, sorting and recycling processes attribute to the low recovery and recycling rates and thus increase the final waste streams. Qualman (2017) alludes that only 5% of plastics used in the package industry are recovered for recycling purposes and one-third

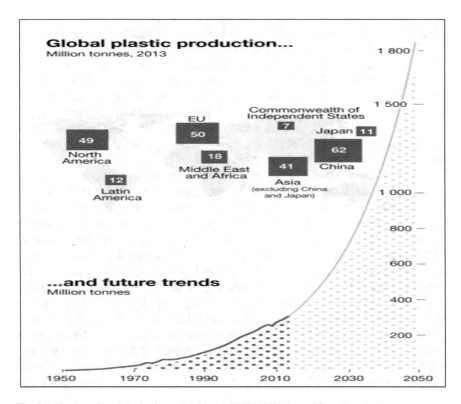

**Fig. 2.1** Projected global plastic production by 2050 ([30] adapted from Ryan)

**Table 2.1** Projected global waste composition by 2025 [14]

*Current estimates*

| Income level | Organic (%) | Paper (%) | Plastic (%) | Glass (%) | Metal (%) | Other (%) |
|---|---|---|---|---|---|---|
| Low income | 62 | 5 | 8 | 3 | 3 | 17 |
| Lower-middle income | 59 | 9 | 12 | 3 | 2 | 15 |
| Upper-middle income | 54 | 14 | 11 | 5 | 3 | 13 |
| High income | 28 | 31 | 11 | 7 | 6 | 17 |

*2025 estimates*

| Income level | Organic (%) | Paper (%) | Plastic (%) | Glass (%) | Metal (%) | Other (%) |
|---|---|---|---|---|---|---|
| Low income | 62 | 6 | 9 | 3 | 3 | 17 |
| Lower-middle income | 55 | 10 | 13 | 4 | 3 | 15 |
| Upper-middle income | 50 | 15 | 12 | 4 | 4 | 15 |
| High income | 28 | 30 | 11 | 7 | 6 | 18 |

escapes inclusion in the garbage collection system. It is illegally disposed of into the environment (i.e. streams, oceans, roadsides and lakes).

The majority of waste categories present containers and packaging plastics with the highest tonnage [3]. For this reason, it is important to establish contextual PSWs management solutions. As a result of a context approach to plastic recycling, in 2012, an increase of 9.3% of PSWs recycling was reported in South Africa. For developed economies in the European Union, a number of strategies have been implemented for the sustainable management of PSWs. A contextual approach is used in designing and implementing most of these solutions [10].

Several research works have reported on how to manage post-consumer PSWs. Most of these studies have focused their attention on developed economies. A study by [5] discussed many technologies for managing plastics from polyolefinic sources. The study paid attention to different types of recycling and energy recovery methods. [15] assessed the challenges and opportunities for managing PSWs. The study identified a number of opportunities and challenges in PSWs recycling. The two studies identified different plastic recycling and energy recovery methods applicable in the context of developing economies. Further, the studies highlighted the solutions for managing PSWs in developed economies. Al-Salem et al. [5] highlighted that to achieve success in the practical applications on the various plastic recycling methods, the end products resulting from different mechanical treatments should have similar properties of commercial-grade plastics that are aligned to the monomer type and origin. This strategy is workable in both developed and developing economies' industries and should be taken into consideration when deciding the type of plastics to recycle. The study by [15] noted that challenges of both technological factors and socioeconomic behavior issues relating to the recovery of recyclable wastes and substitution for virgin materials still exist. The affirmation of [15] implies that efforts and solutions are needed to sustainably recover PSWs. Further, the quantity of post-consumer PSWs is more than the pre-consumer waste by 3.5–5 in factor rating (depends on data year of sourcing and analysis) with a high probability of future waste increases attributed by extended plastic applications.

## 2.2  Post-consumer Plastic Solid Wastes

Post-consumer plastics are plastics that have been produced into a product, recovered and remanufactured into a new product. The packaging industry represents a huge number of products that form post-consumer plastic wastes. Products such as plastic bottles, bags, toys, etc. constitute post-consumer plastic products. Figure 2.2 depicts an example of post-consumer products.

According to the EPA, a post-consumer product is an EoL product or material that is recovered or diverted before it is disposed of. These are the materials that businesses and consumers recycle.

Recovery of post-consumer products contributes to closing the loop, by diverting EoL post-consumer PSWs from the landfills. In developed economies, approximately

**Fig. 2.2** A picture of post-consumer plastic products

50 million tons of post-consumer PSWs are generated [6]. As a result of the establishment of sustainable systems for recovering and recycling PSWs, countries of developed economies are closing the loop. Figure 2.3 depicts the recovery rates for post-consumer plastic films in the USA from 2005 to 2016. The figure shows a progressive increase in the recovery rates over a period of 12 years. For example, in 2016, 1.3 billion pounds of post-consumer PSWs films were recovered. The rate of recovery of post-consumer PSWs films has doubled since 2005. Further, options of

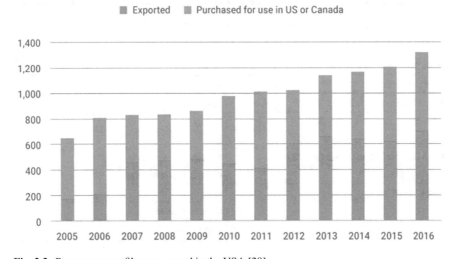

**Fig. 2.3** Post-consumer films recovered in the USA [28]

recovering post-consumer PSWs films for sale to other nations are used. For example, of the 1.3 billion tons of plastic films recovered, 47% was recycled and 53% was exported to other nations.

In developing economies, the recovery of PSWs for recycling and other sustainable management options is still in its infancy [31]. Nevertheless, it is important that strategies and systems for recovering post-consumer PSWs used in developed economies are tested in developing economies. Testing of existing strategies and systems in developing economies will enable waste managers and engineers to apply contextualized strategies and systems.

A number of post-consumer plastic packaging products exist and it is important for waste managers, recyclers and engineers to understand. In Sect. 2.1, two categories of plastics are discussed: thermoplastics and thermosetting plastics. These categories

**Fig. 2.4** Piles of unrecovered PSWs dumpsite [picture captured on January 6, 2020]

**Table 2.2** Categories of plastics

| Thermoplastics | Thermosets |
| --- | --- |
| Polyethylene terephthalate (PET) | Alkyd |
| Low-density polyethylene (LDPE) | Epoxy |
| Polyvinyl chloride (PVC) | Ester |
| High-density polyethylene (HDPE) | Melamine formaldehyde |
| Polypropylene (PP) | Phenolic formaldehyde |
| Polystyrene (PS) | Silicon |
| | Urea formaldehyde |
| | Polyurethane (PU) |
| | Metalized and multilayer plastics |

**Table 2.3** Post-consumer recycling identification codes

| Number code | Abbreviation | Original name | Usages |
|---|---|---|---|
| 1 | PET | Polyethylene terephthalate | Food containers, bottles (beer and soft-drinks) |
| 2 | HDPE | High-density polyethylene | Grocery bags, cosmetic containers, pails |
| 3 | PVC | Polyvinyl chloride | Wire insulation, pipes, tubing |
| 4 | LDPE | Low-density polyethylene | Bags and film |
| 5 | PP | Polypropylene | Drug and food containers |
| 6 | PS | Polystyrene | Expanded foam, plates, food packaging, cutlery, CD holders |
| 7 | Other | | |

of plastics can be either synthetic and semi-synthetic. Table 2.2 depicts the types of plastics under the two categories.

It is important to indicate that most post-consumer plastic packaged products are made from thermoplastics. Based on the definition of post-consumer plastics, the ability of thermoplastics to undergo strong molecular motions, causing them to soften, allows for the manufacture of post-consumer products. Table 2.3 depicts the types of post-consumer recycling identification codes that engineers, waste managers and recyclers should know. An understanding of the identification codes is important for sorting and collection processes. The identification codes are usually used on packaging, bags, bottles and containers.

HDPE and PET are commonly used for consumer product packaging as a result of their availability [27]. In addition, existence of robust markets for the commonly recycled plastics in the industrial sector has contributed to plastic recycling. Over the years, the manufacture of post-consumer packaging products from recycled plastics has presented economic and environmental advantages compared to manufacturing from virgin plastic materials. According to Smith [38], the carbon footprint generated from the manufacture of PET bottles from 100% post-consumer recycled material is 60% lower compared to the utilization of virgin PET plastic material. Further, utilization of PCR material eliminates the consumption of virgin plastic materials and this is critical considering the low recycling rates globally. Smith [38] notes that low recycling rates are attributed to low consumer demand for post-consumer products. However, with the high demand for PSWs, recycling rates can increase.

## 2.3   Sources of Post-consumer Plastic Solid Wastes

In developing and developed economies, the management of PSWs is one of the many challenges that authorities in large and small cities face. This is attributed to the increasing demand for plastic packaged products and the insufficient budgets of the municipality. In addition to the high costs of management, a lack of understanding of the numerous factors that affect the PSW management system also contributes. One factor that requires understanding is the source from which PSWs are generated. This aspect is relevant in the supply chain of management of PSWs. Recoveries of PSWs need to understand where the optimal recoveries are generated. BIO-Intelligence [35] notes that recovery rates are sustainable by closer engagement of the major stake-holders along the supply chain. Manufacturers of plastic products, waste managers and the IWCs should understand the logistics of managing PSWs.

PSWs are generated from a number of sources but the quantities generated at these sources are usually not recorded. A study conducted in 22 developing economies on waste management revealed that few keep quantitative data on the amount generated [13]. Nevertheless, the non-existence of quantitative data should not prevent understanding of the sources where wastes are generated. Different waste streams constitute PSWs; for example, municipal solid wastes; domestic, construction and demolition; and industrial, commercial and institutional. All these waste streams constitute PSWs but the quantities are different. According to [19], the composition of MSWs differs in municipalities and this applies from country to country. Such variation is based on the industrial structure, lifestyles, economic situation and waste management regulations.

Information on the sources of PSWs is a critical driver for determining appropriate management and handling techniques. The fact that 50% of plastic materials are used to manufacture packaging products implies that 50% is generated from plastic pack-ages. A study conducted by Hoornweg and Bhada [14] reveals that the sources of plastic wastes are bottles, containers, bags, lids, packages and cups. PSWs are gener-ated from variable sources as a result of the different applications. MSWs constitute different types of wastes and Fig. 2.3 depicts a dumpsite in Kenya filled with EoL plastic products. In developing economies, the majority of PSWs are recovered from dumpsites because there is a lack of waste recovery systems that are points of gener-ation. A study conducted by Mwanza [49] reveals that majority of the scavengers are found at the dumpsites.

## 2.4   Trends in Managing Post-consumer PSWs
##         in Developing Economies

WM in developing economies encounters a number of challenges. As a result, 59% of the waste generated is disposed of at landfills [14]. However, a number of changes are happening in developing economies to improve the WM sector. The call for SWM is pushing the governments to enforce effective and efficient methods of WM.

Post-consumer PSWs constitute MSWs and this implies that its management is conducted with other wastes such as papers, glass, metal, rubber and other waste components. The properties possessed by plastic materials have favored them among other materials and thus their inclusion in most waste streams. Sustainable options for managing EoL post-consumer PSWs are not limited to the 3Rs.

In developing economies, the majority of wastes are recovered by the informal waste collectors for recycling, reusing and/or remanufacturing purposes. Scheinberg et al. [33] point out that informal recyclers recover appropriately 15–35% of waste generated in cities of developing countries. However, according to Hoornweg and Bhada [14], in Africa, only 4% of MSW is recycled while in Asia 8% is recycled. These percentages are a representation of all recyclable wastes. The percentages indicate that a smaller percentage constitutes the recycling of PSWs. These low percentages are a call for implementing more sustainable ways of recovering PSWs.

Post-consumer PSWs management in developing economies has been driven by public health-related issues. This drive to manage wastes is not different from developed economies. Nevertheless, developed economies are now driven to manage wastes by sustainability-related issues instead of public health. MSW has always been collected to avoid the outbreak of diseases, and in this way, post-consumer PSWs were managed before the realization of their value. Environmental concerns regarding WM are not the main reason for proper management of waste in most developing economies [44] since a majority of wastes are still not collected. According to [14], less than 45% of MSW is collected in Africa and 47% is openly dumped. Therefore, the trend in the management of PSW has not been completely influenced by environmental concerns.

The recovery of PSWs by the informal waste sector (IWS) has been driven by the economic value attached to plastics. The feasibility of a livelihood by recovering tradeable waste has driven most of the urban poor into the recovery chains [44]. The majority of PSWs are recovered for recycling and reusing purposes in developing economies. According to [14], greater than two million informal waste collectors (IWCs) are engaged in the recycling sector. The increase in the amount of post-consumer PSW generations has contributed to the provision of jobs in the IWS. However, the processing of PSWs into other products by the IWS has come with a number of environmental challenges. Zhang and Wen [47] noted that most of the small to medium industries of PET lack proper equipment and pollution control mechanisms. Despite these challenges, the recovery of PSWs is mainly conducted by the IWS. Velis et al. [42] pointed out that considerable recycling rates of between 20 and 30% in low-income countries are achieved through the IWS participation and this has contributed to saving WM expenses by 20% in the local government units. Table 2.4 shows a comparison of material recovery between the formal and informal waste sectors. The table indicates that in most of the cities except Lusaka, the informal sector has dominated the recovery of waste. This trend in the management of waste, in particular PSWs, has also been driven by the livelihood the informal sector gets from recovering this waste. In developing economies of low- and middle-income levels, hundreds to thousands of families depend on recycling entrepreneurial

**Table 2.4** Comparison of waste recovery between the formal and informal sectors [33]

| City | Formal sector | | Informal sector | |
|---|---|---|---|---|
| | Tons | % of Total | Tons | % of Total |
| Cairo, Egypt | 433,200 | 13% | 979,400 | 30% |
| Cluj, Romania | 8900 | 5% | 14,600 | 8% |
| Lima, Peru | 9400 | 0.3% | 529,400 | 19% |
| Lusaka, Zambia | 12,000 | 4% | 5,400 | 2% |
| Pune, India | – | 0% | 117,900 | 22% |
| Quezon City, Philippines | 15,600 | 2% | 141.800 | 23% |

activities for a livelihood and this has contributed to the formation of recycling supply chain pyramids [7, 37].

It should be noted that in low- and middle-income countries, waste recycling is different from WM as it is not a cleaning or waste removal service but an economic activity that is linked to the industrial sector [23, 34]. The link to the industrial sector indicates that after recovery most of the wastes are recycled, reused or remanufactured.

## 2.5 Challenges in Managing Post-consumer PSWs in Developing Economies

Many challenges concerning PSWs management exist in developed and developing economies. It is one of the reasons that Ellen MacArthur has become a global advocate for the circular economy. The 2017 report, "The Latest Plastic Economy: Reaffirming the Destiny of Plastics and Modifying Action," lays out a 50–30–20 plan for plastics. It outlines that 50% of PSWs should be recycled, 30% should be redesigned and 20% should replace reusable alternatives. Despite the existence of the drive for a circular economy, many countries in developing economies are still facing WM problems. A study conducted by [36] outlines many WM challenges faced in developing economies and most of these challenges have been affirmed in other studies [2, 4, 18, 21, 29]. Even though these studies did not completely look at PSWs, Hopewell et al. [15] assessed the challenges and opportunities of recycling PSWs, and it is not surprising that the challenges faced are not in isolation from the challenges faced by MSWs managers. SW managers and engineers should be aware of these challenges as they manage PSWs.

## 2.5.1  Sorting of Post-consumer Plastic Solid Waste

Sorting of post-consumer PSWs is one of the first stages in the recovery process and despite how efficient the recovery process is, sorting is an important stage in the recovery loop. Regardless of many sorting technologies, the removal of paint from plastic products is one of the major problems that recyclers encounter. Plastic properties are normally compromised by stress concentration developed by coating materials [16]. As a result of these challenges, recyclers or waste managers should identify appropriate types of sorting technologies. Manual and automatic techniques are used for sorting co-mingled rigid recyclables. Plastic streams separation from paper, glass, wood or metal is achievable through automated presorting. Majority of material recovery facilities (MRFs) rarely collect post-consumer packaging because of the current deficiencies in the material separation equipment [15]. However, many technologies with the ability to increase the recovery rates for post-consumer PSWs exist and are discussed in Chap. 3.

## 2.5.2  Collection and Segregation of Post-consumer PSWs

A number of developing economies have sufficient waste collection challenges. In a study conducted by [14], it was affirmed that less than 50% of the waste generated is collected for disposal. In a number of developing economies, the municipalities, IWCs and private collectors are actively involved in the segregation and collection of recoverable PSWs. However, the majority of the studies indicate that the IWCs are the major players in the segregation and collection of recoverable PSWs [8, 11, 13, 22, 32]. Waste collection and segregation in developed economies have advanced because of the application of technology. In developing economies, manual collection and segregation continue to exist and this has placed challenges on the recoveries. For example, the IWCs lack structured systems and technology to collect and segregate PSWs, yet they are the major recoveries [11]. Further, households in developing economies continue to dispose of wastes illegally while others use primitive methods such as burning and burying PSWs.

The concern of waste managers and the stakeholders with WM should understand that a major health risk is being caused by wastes accumulating in dumpsites that are near residential areas. Therefore, it is necessary for the formal waste sector (FWS) to integrate the IWS into formalized waste collection systems since they are the major recoveries. Integration of the IWCs into formalized systems will promote sustainable WM and hence a positive drive toward achieving a circular economy.

Source waste segregation is a major challenge in most developing economies. According to [41], a number of waste treatment alternatives are not attainable and economical in the absence of source waste segregation. Structured designed source segregation of wastes including PSWs can create opportunities for scientific disposal of wastes [41]. To decrease the challenges of waste collection and segregation, waste managers, engineers and parties interested in WM should design an enabling environment for source segregation with additional increased processing facilities. Furthermore, the inclusion of the IWCs through selective source segregation is an intermediary step for alleviating collection and segregation challenges in developing countries.

### 2.5.3  Many Waste Collection Points and Quantities

Considering a global perspective on PSWs, waste generation continues to increase and these are the avenues that waste managers should understand and think beyond. A number of studies have affirmed that the amount of SW will continue to increase because of increasing socioeconomic developments and urbanization. Karak et al. [17] pointed out that by 2050 SW generation will increase to 27 billion globally. Hoornweg and Bhada-Tata [14] affirmed that by 2025 MSWs generation in most world cities will be generated. The increase in waste generation does not exclude PSWs and developing economies. Minghua et al. [24] mention similar developments in developing economies. However, in developing economies, the problems of managing PSWs and other wastes are massive compared to developed economies. Furthermore, the number of waste collection points and areas continues to increase because of urbanization and socioeconomic changes. For those reasons, waste managers should understand and design appropriate strategies for managing PSWs amid the existing challenges of increased areas to serve and increased quantities of waste to collect.

For developing economies, incorporation of the IWCs into formalized systems is a strategy that can work in bridging the gap between increased PSWs generation and the number of areas requiring service. For example, a number of studies have developed frameworks for integrating the IWCs into formalized systems [42, 46]. These studies have developed these frameworks for developing economies. Furthermore, Matter et al. [21] and Storey [39] point out that source segregation at the household level is required to enable sustainable waste collection and recovery.

### 2.5.4  Insufficient Resources

SWM receives low priority in the local government budget and ends up with inadequate funds. Sometimes there is a common budget for treatment and collection. The evidence of inadequate budgets is seen through illegally disposed of and uncollected PSWs in urban areas. The increase in illegally disposed of PSWs in urban areas is because of the increase in population growth as people continue to search for better opportunities. With inadequate resources for waste collection and treatment, many developing economies have not invested in plastic treatment options such as recycling and energy recovery. Most of the post-consumer packaging wastes are recovered by the IWS with limited knowledge and facilities, to adequately and properly recycle these PSWs. According to a study by [47], 90% of PET bottles are recovered by IWCs and reprocessed in small factories without proper pollution control mechanisms. This shows that insufficient resources are allocated to manage and support the informal and formal sectors in PSWs recovery business ventures. The inadequacy of trained IWCs is also another challenge faced in developing economies and can be attributed to insufficient funding to WM activities.

### 2.5.5 Management and Societal Apathy

The operational efficiency of SWM depends on the active involvement of citizens and the municipal agency. The fact that the social status of SWM is low, apathy toward it exists. Signs of apathy include uncollected waste, deterioration of aesthetics in many areas and reduced environmental quality because of uncontrolled disposal sites.

Many societies attribute WM to the responsibility of the local authorities. Wastes are illegally disposed of and littered in the expectation of the local authorities to clean up. A number of communities have no waste disposal systems and end up disposing of their wastes illegally at night. They have developed a societal apathy toward WM and continue to blame the local authorities for uncollected wastes.

Societal and management apathy toward waste should be resolved by integrating the key stakeholders in the WM arena. These stakeholders should be integrated with clearly designed responsibilities for driving SWM. For example, a study conducted on the factors enhancing extended producer responsibility (EPR) revealed that the responsibilities of the WM companies (public and private) and the manufacturers are categorized into physical and financial responsibilities (Mwanza and Mbohwa 50) and this has contributed to the successful enforcement of EPR in developed economies. The WM companies are responsible for ensuring proper collection and disposal of waste while the manufacturers are responsible for ensuring the waste companies are funded. Therefore, the waste managers and engineers should develop systems that should place waste management responsibilities on everyone in order to end the apathies.

### 2.5.6 Household Education

In all aspects of life, education is considered important. Public knowledge on WM and understanding of the connection of human behavior, sanitation and waste handling require household education. However, in most developing economies, household education on WM is still in its infancy. Household education should be an enabling factor for driving SWM. Thus, in most studies, provision of knowledge and awareness on PSWs is affirmed as an enabling factor to sustainable resource utilization and WM [1, 25, 43, 47] (Nixon and Saphores 51). To overcome the barrier between

SWM and household education, waste managers, engineers and educators should identify the best strategies for ensuring messages on sustainable PSWs management are disseminated to the public. For example, in Jamaica, the waste administrators implemented public education campaigns as strategies for alleviating the rampant littering behavior of its citizens [26].

Education is a crucial solution for increasing awareness and behavioral change toward PSWs management. It is the future solution for reducing the indiscriminate disposal of PSWs by members of society. Most studies present solutions such as recycling, reusing, incinerating etc. as sustainable strategies for managing PSWs but the aspect of community or household education on PSWs is usually not mentioned. For this reason, the approach of educating the communities on the dangers of poor PSWs management should be encouraged and enforced. This can be conducted through anti-littering campaigns and promotions on sustainable ways of managing PSWs. For example, in Zambia, the last day of every month was declared a LET'S KEEP ZAMBIA CLEAN CAMPAIGN by his Excellency the President. Members of the public are encouraged to participate in cleaning the city by sweeping and picking up littered wastes. This has become a way of life and it keeps promoting the drive toward zero waste. The companies have engaged in the same noble cause by participating in this campaign and this is reflected as one of their corporate social responsibility (CSR) activities.

The local government, private waste collecting companies, the public and the corporate world can engage in the memorandum of agreements on how to educate the communities on sustainable ways of managing wastes. This should be conducted in order to help the local governments and the organizations mandated to manage wastes. Further, the local authorities and the Ministry of Education should work together to establish sustainable ways of educating the future generation on the criticality of PSWs management. One way is incorporating in the school curriculum topics on SWM.

In addition, household education on the management of wastes should not be limited to sustainable strategies such as reuse, recycle and recover, but the education should address aspects of waste regulations and rules at the community level.

# References

1. R. Afroz, A. Rahman, M.M. Masud, R. Akhtar, The knowledge, awareness, attitude and motivational analysis of plastic waste and household perspective in Malaysia. Environ. Sci. Pollut. Res. **24**, 2304–2315 (2017)
2. O.N. Agdag, Comparison of old and new municipal solid waste management systems in Denizli, Turkey. Waste Manage. **29**, 456–464 (2008)
3. Agency, U.S. Environmental Protection, *EPA's 2008 Report on the Environment*. National Center for Environmental. Assessment (2008)
4. J.M. Alhumoud, F.A. Al-Kandari, Analysis and overview of industrial solid waste management in Kuwait. Manage. Environ. Qual. **19**(5), 520–532 (2008)
5. S.A. Al-Salem, P. Lettieri, J. Baeyens, Recycling and recovery routes of plastic solid waste (PSW): a review. Waste Manage. **29**, 2625–2643 (2009)

6. J. Baeyens, A. Brems, R. Dewil, Recovery and recycling of post-consumer waste materials—Part 2. Target wastes (glass beverage bottles, plastics, scrap metal and steel cans, end-of-life tyres, batteries and house hold hazardous waste). Int. J. Sustain. Eng. **3**, 232–245 (2010)

7. A.R. Chaturvedi, R. Kilguss, U. Arora, E-waste recycling in India—bridging the formal–informal divide, in *Environmental Scenario in India: Successes and Predicaments*, ed. by Mukherjee, D. Chakraborty (Routledge, London, 2011)

8. T.M. Coelho, PET containers in Brazil: opportunities and challenges of a logistics model for post-consumer waste recycling. Resour. Conserv. Recycl. **3**(55), 291–299 (2011)

9. Consumption, British Plastic Federation Industry 2008 Oil, Accessed May 20, 2016 (2008). http://www.bpf.co.uk/Oil-Consumption.aspx

10. N.F. Cruz, S. Ferreira, M. Cabral, P. Simões, R.C. Marques, Packaging waste recyclingin Europe: is the industry paying for it? Waste Manage. **34**, 298–308 (2014)

11. C. Ezeah, J.A. Fazakerley, C.L. Roberts, Emerging trends in informal sector recycling in developing and transition countries. Waste Manage. **33**, 2509–2519 (2013)

12. L.A. Guerrero, G. Maas, W. Hogland, Solid waste management challenges for cities in developing countries. J. Waste Manage. **33**, 230–232 (2012)

13. J. Gutberlet, *Recovering Resources—Recycling Citizenship: Urban Poverty Reduction in Latin America* (Ashgate, Aldershot, 2008)

14. D. Hoornweg, P. Bhada-Tata, *What a Waste: A Global Review OfSolid Waste Management. Urban Development Series Knowledge Papers* (World Bank, Washington, DC, 2012)

15. J. Hopewell, R. Dvorak, E. Kosior, Plastics recycling: challenges and opportunities. Phil. Trans. R. Soc. B **364**, 2115–2126 (2009)

16. H.-Y. Kang, J.M. Schoenung, Electronic waste recycling: a review of US infrastructure and technology options. Resour. Conserv. Recycl. **45**, 368–400 (2005)

17. T. Karak, R.M. Bhagat, P. Bhattacharyya, Municipal solid waste generation, composition, and management: the world scenario. Crit. Rev. Environ. Sci. Technol. **42**(15), 1509–1630 (2012)

18. N. Khalil, M. Khan, A case of municipal solid waste management system for a medium-sized Indian City, Aligarh. Manage. Environ. Qual. Int. J. **20**(2), 121–141 (2009)

19. M.S.M. Hussein, I.A.S. Mansour, Solid waste issue: Sources, composition, disposal, recycling, and valorization. Egypt. J. Petrol. **27**, 1275–1290 (2018)

20. A.L. Manaf, M.A. Samah, I.M.N. Zukki, Municipal solid waste management in Malaysia: practices and challenges. Waste Manage. **29**, 2902–2906 (2009)

21. A. Matter, M. Dietschi, C. Zurbrügg, Improving the informal recycling sector through segregation of waste in the household–The case of Dhaka Bangladesh. Habitat Int. 150–156 (2013)

22. M. Medina, *The World's Scavengers: Salvaging for Sustainable Consumption and Production* (AltaMira Press, 2007)

23. M.V. Melosi, *Garbage in the Cities: Refuse, Reform, and the Environment* 1 (A&M University Press, College Station, Tex, Texas, 1981)

24. Z. Minghwa, F. Xiumin, A. Roveta, H. Qichang, F. Vicentini, L. Bingkai, H. Giusti, L. Yi, Municipal solid waste management in Pudong New Area, China. Waste Manage. **29**, 1227–1233 (2009)

25. A. Omran, A. Mahmood, H. Abdul Aziz, G.M. Robinson, Investigating households attitude towards recycling of solid waste in Malaysia: a case study. Int. J. Environ. Res. **3**(2), 275–288 (2009)

26. P.S. Pendley, *Feasibility and Action Plan for Composting Operation Incorporating Appropriate Technology at Riverton Disposal Site, Kingston, Jamaica* (Master Of Science In Environmental Engineering, Kingston, 2005)

27. PlasticsEurope, *An analysis of European plastics production, demand and waste data* (2018). May 20. Accessed 2019.

28. Recycling, MORE, *2016 National PostConsumer Plastic Bag and Film Recycling* (2018). Accessed May 12, 2019

29. S. Mohsen,A framework for sustainable waste management: challenges and opportunities. Manage. Res. Rev. **38** (2015)

30. P.G. Ryan, A brief history of marine litter research, in *Marine Anthropogenic Litter* ed by M. Bergmann, L. Gutow, Klages (Springer, Berlin, 2015)
31. J. Sarkis, M.M. Helms, A.A. Hervani, Reverse logistics and social sustainability. Corp. Soc. Responsibility Environ. Manag **17**(6), 337–354 (2010)
32. A. Scheinberg, *Value Added, Modes of Sustainable Recycling in the Modernisation of Waste Management Systems.* PhD dissertation,, Wageningen University, The Netherlands. Gouda.The Netherlands: WASTE (2011)
33. A. Scheinberg, M. Simpson, Y. Gupt, *Economic Aspects of the Informal Sector in Solid Waste Management* (GTZ (German Technical Cooperation) and the Collaborative Working Group on Solid Waste Management in Low and Middle Income Countries (CWG), Eschborn, Germany, 2010)
34. A. Scheinberg, The proof of the pudding: Urban recycling in North America as a process of ecological modernisation. Environ. Polit. **12**(4), 49–75 (2003)
35. Service, BIO Intelligence, Study on an increased mechanical recycling target for plastics Recyclers Europe Recyclers Europe. Final report prepared for Plastic (2013)
36. V.A. Shekdar, Sustainable solid waste management: an integrated approach for Asian countries. Waste Manage. **29**, 1438–1448 (2009)
37. M. Simpson-Hebert, A. Mitrovic, G. Zajic, M. Petrovic, *Belgrade's Roma in the Underworld of Waste Scavenging and Recycling* (WEDC, Loughborough University, UK, Leicestershire, 2005)
38. K.T. Smith, *The Benefits of Using Post-Consumer Plastics* (Bath, UK, 2019)
39. D. Storey, L. Santucci, R. Fraser, J. Aleluia, L. Chomchuen, Designing effective partnerships for waste-to-resource initiatives: lessons learned from developing countries. Waste Manage. Res. **33**(12), 1066–1075 (2015)
40. Trends, Global Business, *How Plastics Waste Recycling Could Transform the Chemical Industry* (2015)
41. UNEP, *State of the Environment* (2001). Accessed 2016. http://www.eapap.unep.org/reports/soe/
42. C.A. Velis, D.C. Wilson, O. Rocca, S.R. Smith, A. Mavropoulos, C.R. Cheeseman, An analytical framework and tool ('InteRa') for integrating the informal recycling sector in waste and resource management systems in developing countries. Waste Manage. Res. **30**(9), 43–66 (2012)
43. P. Vicente, E. Reis, Factors influencing households' participation in recycling. Waste Manage. Res. **26**, 140–146 (2008)
44. D.C. Wilson, C. Velis, C. Cheeseman, Role of informal sector recycling in waste management in the developing countries. Habitat Int. **30**, 787–808 (2006)
45. D.C. Wilson, C.A. Velis, L. Rodic, Integrated sustainable waste management in developing countries Proc. Inst. Civil Eng. Waste Res. Manage **166**, 52–68 (2013)
46. D. Xevgenos, C. Papadaskalopoulou, V. Panaretou, K. Moustakas, D. Malami, Success stories for recycling of MSW at municipal level. Waste Biomass Valor **6**: 657–684
47. H. Zhang, Z.G. Wen, The consumption and recycling collection system of PET bottles a case study of Beijing, China. Waste Manage. **34**, 987–998 (2014)
48. D. Qualman, Global plastic production 1917 to 2017. https://www.darrinqualman.com/global-plastics-production/global-plastic-production-1917-to-2017/ Accessed 12 March 2020 (2017)
49. B.G. Mwanza, An African reverse logistics model for plastic solid wastes, Ph.D. (Engineering Management): University of Johannesburg (South Africa). ProQuest Dissertations Publishing, 28284643 (2018)
50. B.G. Mwanza, C. Mbohwa, Strategies for enhancing extended producer responsibility enforcement: a review, in *Procs West Africa Built Environment Research (WABER) Conference*, ed. by S. Laryea, E. Essah (Accra, Ghana, 2019), pp. 921–930
51. H. Nixon, J-D.M. Saphores, Information and the decision to recycle: results from a survey of US households, J. Environ. Plan. Manage. **52**(2), 257–277 (2009). https://doi.org/10.1080/09640560802666610

# Chapter 3
# A Review of Technologies for Managing Plastic Solid Wastes

This chapter discusses the different technologies for managing PSWs. Literature works on the different technologies are reviewed with the aim of establishing sustainable technologies for developing economies. Different technological methods such as composting, incineration, recycling and landfilling are presented. The discussion on the technological methods for treating PSWs focuses on the economic and environmental aspects. Furthermore, current trends in each technology are discussed. The chapter provides insights to waste managers and policy makers in developing economies.

## 3.1 Recycling of Plastics

Recycling re-introduces parts of energy and materials that would have been considered waste back into the system [9]. Manzini [21] indicate that recycling extends the lifespan of a products' material more than its products' initial lifespan. Recycling can either be an open-loop or closed-loop process. Pacheco [28] adds that recycling occurs from the stages of sorting, collection and processing.

*Open-loop recycling involves the conversion of a product's material into a different product type [21]. This process applies to open systems in which the material recycled into a different product undergoes changes in its inherent properties [5].*

*Closed-loop recycling involves the conversion of a byproduct of a product into another product [21]. This procedure applies to systems in which the recycled material inherent properties do not change [5]. However, [16]add that the practicality of closed-loop recycling is effective when the sources of contamination are separated*

B. G. Mwanza and C. Mbohwa, *Sustainable Technologies and Drivers for Managing Plastic Solid Waste in Developing Economies*, SpringerBriefs in Applied Sciences and Technology, https://doi.org/10.1007/978-3-030-88644-8_3

*from the polymer constituent and if the degradation is stabilized during subsequent use and reprocessing.*

Recycling is one of the 3Rs providing a platform for reducing carbon dioxide emissions, amounts of waste disposals and oil usage. However, one can wonder about the impact that waste prevention has on the environment as being the first R on the waste hierarchy. WFD alludes that "prevention" is the measure or action taken before a product, substance or material becomes a waste. The question that arises is how to measure what is not produced [5]. Waste reuse holds a second position on the waste hierarchy followed by recycling. Nevertheless, WFD shows that reuse is any action taken in which a waste product is used for the purposes it was initially manufactured to perform [5]. In the case of waste, recycling has therefore been considered as a process currently sustainable in providing the opportunity to decrease the amounts of oil usage, waste disposed and carbon dioxide emissions [16]. Fundamentally, increased recycling rates allow for product service with reduced material inputs than would be required provided a reduction in use, remanufacturing or reuse [16]. Other than that, recycling is a WM strategy, which is an example of the industrial ecology concept implementation even though in the natural ecosystem wastes do not exist [12, 23]. From the fact that recycling is a WM strategy, it is necessary for waste managers, engineers and recyclers to understand the various recycling technologies available.

In developing economies, the majority of wastes are recovered by the IWS, and in a survey conducted by [45], the majority of the PET wastes are reprocessed in factories designed without pollution control equipment. Understanding the various types of technologies for recycling wastes and paying attention to the economic and environmental aspects are necessary for decreasing environmental pollution in developing economies. Al-Salem et al. [2] affirm that plastic recycling is cardinal for many reasons such as oil conservation, greenhouse gas emission reduction, landfill space preservation, energy conservation and reuse benefits. As a result of these important plastic recycling benefits, most developed economies have increased environmental awareness and legislative measures as a road map to sustainable recycling.

In developed economies, i.e. Germany, The Netherlands, Japan, the USA and China, PSWs recycling is legalized. For example, China imports PSWs from other nations for recycling purposes. In 2016, approximately one-third of the world's PSWs were imported by China for recycling. For Japan, a number of drivers have contributed to it, becoming the world's highest recycler. According to PWMI, Japan achieved a recycling rate of 77%. Despite having the highest recycling rates from a number of developed nations, approximately 8 million metric tons of PSWs end up in the oceans annually [13]. It is anticipated that, by 2050, the oceans will occupy more plastics than fish [27]. These are healthy and environmental implications of unsustainable management of PSWs. This necessitates the agency in discussing and addressing the different types of technology and systems available to prevent unsustainable management of PSWs.

In developing economies, PSWs recycling rates cannot be comparable to developed economies. However, in developing economies, PSWs recycling is happening through the IWS. Making a living out of waste recycling and collection motivates the majority of the IWS to engage. Nevertheless, it is difficult to quantify the amount of PSWs collected and recycled because the majority of the IWS lacks structured MRFs and systems. From the fact that the majority of PSWs recovery is conducted by the IWS with a few private-owned companies, it is necessary to create value chains systems. A proposition for designing these systems is for the interested parties to understand the existing recycling technologies. These technologies should be understood from a contextual perspective.

### 3.1.1 Understanding Plastic Recycling

The recovery and recycling of plastics require an understanding of the different technologies available. According to [16], plastics recycling is a complex and confusing terminology because of the variety of activities involved. As a result, SW managers, engineers or parties involved or interested in starting up MRFs and recycling centers need to understand the different types of recycling technologies. Generally, there are four categories: primary, secondary, tertiary and quaternary recycling technologies. Mastellone [22] asserts that PSWs treatment and recycling processes are defined under four categories: energy recovery (quaternary), mechanical (secondary), chemical (tertiary) and re-extrusion (primary). Despite the differences in application and technology, each category has unique advantages that are beneficial for specific requirements and locations [2].

Another important aspect for consideration during the selection of PSWs recycling technology is whether the technology decreases the requirements for new inputs of resources or limits environmental burdens. It is cardinal to choose plastic recycling methods with the least social costs and limited environmental impacts while taking into consideration the types of PSWs to be recycled. Other aspects to understand in PSWs recycling are recycling costs, recovery costs, availability of recyclable PSWs, integrated participation of key stakeholders (i.e. households, IWS, private waste sector and the municipality) and government involvement through regulations and legislations.

Understanding the recycling costs of PSWs is the key to determining whether recycling-specific plastic types are sustainable and feasible. For PSWs recyclers, it is necessary to understand the costs of recycling before implementing the recovery and recycling project. Recycling and recovery of many PSWs have failed due to lack of understanding of the recycling costs involved. The recovery costs include logistic costs as such movement of PSWs from the point of generation to the MRFs. Logistic costs include transportation costs, administration costs, rent, customer service, labor, inventory carrying, supplies and others. Availability of recyclable PSWs is important for successful recycling projects. A number of projects fail due to lack of recyclable PSWs. Further, understanding the recyclability of PSWs is relevant to the type of recycling technology. Sustainable recovery and recycling are most effective and successful with the inclusion of the key stakeholders in the value addition chains. Therefore, consideration of these aspects is important for the recycling of PSWs.

## 3.1.2  Primary Recycling

Primary recycling (re-extrusion) is the re-introduction of single-polymer plastic parts and edges or scrap into the extrusion cycle to produce components with similar material properties [2]. This method of recycling is equivalent to closed-loop recycling in that the product or input material is reprocessed into a product or component with similar properties. Al-Salem et al. [2] assert that plastic scrap with similar properties or features to the original product is utilized in the process. This recycling method re-extrudes post-consumer plastics. A typical example of primary recycling that has been conducted under closed-loop conditions involves HDPE milk bottles and clear PET bottles in the United Kingdom (UK) [16].

A number of recycling methods exist and injection molding is a valid example of primary recycling [4]. From the fact that post-consumer wastes constitute five times more of the PSWs generated in commerce and industry, the most important step is to increase collection and recycling in order to increase recycling rates. Households generate more post-consumer waste and in order to make post-consumer primary recycling sustainable, households need to be incorporated in the collection and segregation systems.

Primary recycling centers can produce similar or dissimilar products. For example, a post-consumer end-of-life plastic bottle can actually be recycled into a similar product. Therefore, recyclers should understand the processes involved. Five dissimilar stages are involved in primary recycling. Sorting is the first stage in any recovery and recycling system. Al-Salem et al. [2] affirm that sorting is critical for ensuring the successful recycling of plastics. PSWs can be sorted according to color, type of polymer and purity. Sorting facilitates the shredding process and it should be conducted in the shortest time period in order to affect the recycler's finances positively. To this effect, different types of sorting technologies that can remove labels and adhesives have been developed. For example, density sorting is used, though it does not provide a better solution because of close densities among most plastics. Triboelectric separation is another practical technique for sorting PSWs. It distinguishes different resins by using a rubbing technique. The second stage in the primary recycling process for PSWs is washing the sorted PSWs. This stage facilitates the removal of impurities such as adhesives and labels. Al-Salem et al. [2] indicate that labels and adhesives complicate the recycling process. Therefore, washing of PSWs enhances the quality of the finished product as most of the adhesives and labels are removed. The third stage is the shredding of the washed PSWs. Different conveyor belts deliver the PSWs on different shredders where it is cut into smaller pellets in preparation for recycling into other products. Identification and classification is the fourth stage of primary recycling. At this stage, the shredded plastic pellets are subjected to standardized testing in order to verify the class and quality. Finally, extrusion of the tested and shredded plastic pellets into different plastic products is conducted.

### 3.1.3 Mechanical Recycling

Mechanical recycling involves PSWs recovery for the purposes of reusing it in the manufacturing of plastic products via mechanical methods [22]. The methodology facilitates the production of new products from unmodified PSWs. Sometimes, downgrading is a terminology used when recovered plastics are recycled under conditions that do not involve the utilization of virgin polymer (ASTM Standard D5033). Navarro et al. indicate that mechanical recycling involves depolymerization via glycolysis, hydrolysis or methanolysis to produce a variety of monomers. It is a widely commercialized and promoted technology worldwide that is only performed on single-polymer plastics such as PE, PP, PS etc. Mechanical recycling previously constituted the recycling of industrial PSWS but changes in the technology enable the recycling of PSWs from offices, stores and households (Introduction to plastic recycling 2009). Industrial PSWs generated from distribution, manufacturing and processing activities have suitable raw material properties because of limited impurities and dirt, the ability of the various types of reins to separate and large generation rates [2]. Furthermore, many products in our daily lives such as doors, windows, gutters, bags, pipes are manufactured from mechanical recycling technologies.

Despite quality issues when dealing with mechanical recycling, viable and economic routes are opened for PSWs recoveries, especially for rigid and foam plastics [46]. These opportunities in the recycling industries and WM sector are relevant for waste managers or engineers to take advantage and expand their recovery and collection systems.

Mechanical recycling consists of a number of stages, and Fig. 3.1 depicts a schematic diagram of the processes involved [3].

Al-Salem et al. [2] allude that the majority of plastic products, such as profiles for doors and grocery bags, are produced through mechanical recycling processes.

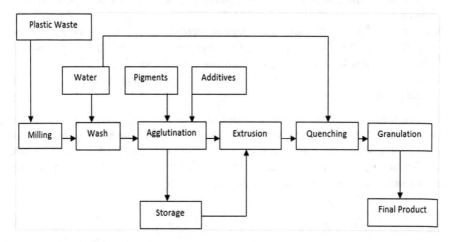

**Fig. 3.1** Mechanical recycling stages [3]

Therefore, it is imperative for recyclers and waste managers to understand the stages involved in mechanically recycling PSWs. In Fig. 3.1, Aznar et al. [3] show schematically the stages involved in mechanical recycling. According to [46] size reduction is the first stage in mechanical recycling. The PSWs are reduced to more suitable pellet powder to flakes. Milling, shredding or grinding are the processes that can be used to reduce the PSWs mechanically. BIO-Intelligence [6] indicates that the efficiency of the process varies for a given amount of PSWs but appropriately 60% can be used. In mechanical recycling, the major recycling stages are discussed below [1, 3, 6].

- Cutting/shredding: At this stage, large PSWs are cut by shearing or sawing into smaller flakes.
- Contaminant separation: BIO-Intelligence (2013) indicates that the collection and sorting of PSWs contribute to the efficiency of recycling lines. However, no matter the efficiency of the recycling scheme, most recyclers face the sorting challenge. For this reason, contaminant separation is designed to remove dust, labels and other forms of impurities from the smaller plastic flakes. Most contaminant separation processes utilize a cyclone.
- Floating: At this stage, various PSWs are separated using a floating tank. The density of the PSWs is used to facilitate the separation of PSWs.
- Milling: In most recycling processes, milling is considered the first stage. However, in mechanical recycling, it is the fourth stage and involves the milling of separated single polymers together.
- Washing and drying: At this stage, pre-washing of the milled polymer is executed using water. Chemical washing is employed for removing glue substances from the milled polymers. Surfactants and caustic are utilized in chemical washing.
- Agglutination: At this stage, pigments and additives are added to the polymer pellets. On completion of this process, the products are either stored or sold. In other cases, the products are further processed through extrusion or a suitable process.
- Extrusion: Through a mold that operates on a screw, melted plastic pellets are continuously extruded in the form of strands. At this stage, a number of products are formed and include pipes, sheets, wire coverings and films.
- Quenching: Using water, plastics are cooled in order to granulate and sell it as a final product.

Once waste managers, recyclers or engineers understand the stages involved in the mechanical recycling of PSWs, the next stage is to understand the specifications for mechanically recycling the different types of recyclable PSWs. Research has shown that a number of authors have expressed interest in the mechanical recycling of polyolefins that end up in PSWs streams [7, 19, 25].

### 3.1.4 Chemical Recycling

The other name for chemical recycling is tertiary. These are advanced forms of technology-based processes that convert PSWs into smaller molecules. The molecules are usually converted into liquids (or gases) that are utilized as feedstock to manufacture new plastics and petrochemicals [22]. PSWs are contributing to the production of valuable fuels and chemicals using chemical recycling technologies. The interest in the production of petrochemicals using PET, nylon and PUR has grown over the past decades. High hydrocarbon content possessed by PET, PUR and nylons has contributed to their increased usage as feedstock [2].

As a result of the chemical alterations of the chemical structure of the PSWs, the term chemical is usually referred to. In developed economies, technological advancement has resulted in most converting companies implementing chemical recycling processes [6]. Depolymerization is the technology that has contributed to the success of chemical recycling, and this has resulted in profitable and sustainable industrial schemes. High product yields and reduced wastes are achievable in chemical recycling.

A number of chemical recycling processes include gasification, pyrolysis, catalyst (or steam cracking), liquid–gas hydrogenation, viscosity breaking and application of PSWs in blast furnaces as a reducing agent [2]. These various types of chemical recycling technologies are suitable methods for preserving scarce resources and environmental protection through decreased volumes of waste. Chemical recycling of PSWs applies sustainable development principles and expectations and follows the principles of sustainable development. The principles of sustainable development are complemented by achieving a circular economy. Thus, for managers, engineers and policy makers, to achieve a circular economy, the application of design for recycling should be supported. Design for recycling requires close collaboration between the product designers and the materials. This implies that a harmonization tool among waste managers, engineers and recyclers is needed.

In developing economies, many challenges on the implementation of advanced recycling technologies are faced. Nevertheless, it should be stressed that advanced recycling technologies contribute significantly to the reduction of virgin material utilization as well as waste reduction. For example, chemically recycled plastics are suitable for application in the food industry and this has increased the application of chemolysis processes. Waste managers and engineers should understand that chemical recycling has broadened the new pathways for manufacturing original value-added products for commercial and industrial utilization. Nevertheless, it should be pointed out that chemically recycled polymers are not as cheap as virgin materials. This is attributed to capital investment, the scale of operation and raw material costs. Managers, engineers and recyclers should understand that to return a profitable economical PET chemolysis facility, approximately 1.5–10.4 tons per annum are needed.

## 3.1.5  Economic and Environmental Impact of Recycling in Developing Economies

PSWs recycling has a variety of economic impacts. To recyclers and manufacturers, recycling is a source of income. In developing economies where WM is a challenge, the establishment of MRFs should be considered. For example, in some developed economies, PSWs are mandatory and landfill usage costs have been made comparable to recycling costs and this has contributed to increased recycling rates.

In densely populated cities, PSWs recycling has the potential to reduce municipal budgets. Some economic analysis has demonstrated that recycling generates three times more in revenue per ton compared to landfilling while more than six times jobs are generated. In a developed economy of St. Louis area, an estimated amount of 16,000 jobs are created, while more than $4 billion yearly revenue is generated. Recycling employs all levels of skilled workers in many sustainable jobs such as material handling, processing and production of high-quality products. In developing economies, the recycling of PSWs is mainly performed by the IWCs. These are motivated by the economic incentive and livelihood they get from recycling activities. Efficient handling and utilization of recycled products and materials contribute to innovation and long-term economic growth. For developing economies, investment in recycling machinery and companies contributes to economic growth, thus filtering the economy.

From an environmental perspective, recycling has positive impacts. It promotes sustainable utilization of natural resources and this has a great impact on the circular economy drive. For example, in developed economies where landfill fees are comparable to recycling costs, this has resulted in the reduction of new landfills, thus saving energy, preventing and reducing pollution, and greenhouse gases emissions. For example, less energy is required to manufacture a new plastic bottle using recycled plastic as recycled plastic has a lower melting temperature compared to virgin raw materials. In essence, recycling prolongs the life of the equipment used to create the plastic product.

### 3.1.5.1  Benefits of Plastic Recycling

One of the aspects of achieving a circular economy is through recycling. It is included in the circular economy because of the number of positive attributes it provides. Because of the fact that PSWs are a major contributor to global waste and cause serious environmental degradation, recycling bridges the gap. In most developing economies plastic bags, bottles and other such products are seen floating in the lakes, rivers and oceans. In addition, PSWs are seen in all regions of human inhabitation. For these and many other demerits of plastics, it is necessary for waste managers, engineers and persons interested in WM and sustainable utilization of

PSWs to understand the benefits of plastic recycling. An understanding of the benefits of plastic recycling influences stakeholders' participation in the establishment and development of MRFs.

Plastic recycling contributes to the conservation of scarce petroleum. Hopewell et al. [16] affirm that almost 4% of energy is required to manufacture plastics, and petroleum is the fuel used in the production of energy. Therefore, the implementation of recycling in the production of products saves approximately 40% of petroleum consumption and this is a huge benefit in the plastic industry.

Greenhouse emissions reduction is another benefit of plastic recycling. Greenhouse gases are generated during the combustion of petroleum. This implies that a reduction in the amount of petroleum utilized during the production of plastics results in a direct reduction in the amount of greenhouse emitted.

Landfill space conservation is again another benefit derived from recycling PSWs. Some studies have indicated that recycling one ton of PSWs spares landfill space of approximately 7.5 cubic yards. In developed economies, landfill fees comparison to recycling costs has resulted in reduced landfill space utilization.

Ingestion of PSWs by animals and birds has caused a number of deaths. A number of aquatic creatures have died from plastic ingestion and recycling reduces the amount of PSWs consumption. From the environmental perspective, recycling reduces deaths in animals and birds. Other than the reduction in the number of animals and birds related deaths, environmental awareness is spread through recycling. Sustainable and judicial use of resources and the creation of green jobs are yet the benefits of recycling.

Global warming is a trending effect because of the continuous utilization of resources and lack of environmental protection. Recycling has mitigated global warming through reduction in pollution and harmful greenhouse gases emission.

From the stakeholders' perspective, especially households, money can be saved through participation in recycling programs. For example, in developed economies, waste collected is charged based on the amount generated. Therefore, the separation of recyclable PSWs reduces the amount collected for disposal, hence reducing the amount paid.

### 3.1.5.2 Challenges in Plastic Recycling

Waste managers and engineers should understand that approximately 10,000 types of plastics are used in the manufacture of different types of products, including toys and even clothes. As a result of these increased developments and innovations in the plastic industry, there has been an increase in the amount of PSWs generated annually. Studies have shown that a large amount of global waste consists of a variety of plastics. This has created a negative impact on the environment especially in developing economies where the WM sector is faced with a number of challenges including lack of waste collection, negative community perception, unenforced legislations and regulations.

In developing economies, many plastic recycling challenges stream from sorting, collection, community engagement, recycling and sorting machinery, source segregation, enforcement of legislations and regulations. Mwanza [26] conducted a study that affirmed that many factors, including incentives provision, provision of waste collection receptacles, mandatory recycling and environmental concerns, influence households' participation in recycling programs. Despite these factors been outlined, a number of communities and companies face a lot of challenges to recycle PSWs. Source segregation is a factor that has not been implemented in most developing economies and households do not engage in source segregation as they believe it is the responsibility of the municipality and the recyclers to sort wastes accordingly. From the recycler's perspective, the variety of plastics, as well as the lack of sorting equipment, has affected recycling negatively. Most recyclers in developing economies are informal and lack sorting machinery.

Illegal disposal of PSWs is evident in many developing economies. Plastics are seen on the streets of most cities and this has given a negative perception. Other than that, a number of diseases such as cholera continue to emerge. The main challenge that engineers, recyclers and managers should understand is the prolific and ever-increasing manufacturing of plastics. This is choking the oceans, planet and our bodies. Further, it should be understood that recyclers have not created the PSWs crisis but the plastic industry continues to create the challenges being faced. PSWs recycling is mediocre in most developing economies, including some developed economies, and this is attributed to the fact that recycling cannot keep up with the expansion of plastics on the market and this is likely to continue. For example, in most developing economies plastic packed products are not locally manufactured and this places a big challenge on the recycling industry. First, the plastic products are imported into the country and extended producer responsibility (EPR) is not effectively enforced, and secondly, weak legislations and regulations on PSWs recyclers mean the distributors are held unaccountable for the amount of PSWs imported and generated into the country. This places a challenge on the recycling industry and on the society as a whole.

Managers, engineers and recyclers should understand that the utilization of plastic materials has increased 20-fold in the last 50 years and only 90% of plastics produced have been recycled since 1950. The majority of the plastics that were recycled were not kept in a continuous loop and were down-cycled into poor grade and single-use products. This challenge continues in most developing economies. According to [42], there is a need for a closed-loop strategy to manage wastes in developing economies. The closed-loop concept of managing and recycling wastes results in an integrated and sustainable utilization of resources.

It is important to acknowledge that plastics have contributed to different sectors of the economy including medicine, transportation and technology. However, from the climatic, health and environmental perspective, the gravity of the challenges PSWs continue to place on the planet is huge to turn a blind eye.

Many challenges exist in the recycling industry, according to the study that was conducted by [16] on the challenges and opportunities of recycling PSWs. The study highlighted many challenges to PSWs recycling. Mixed recycling of PSWs is a major challenge in the recycling industry. The majority of the generators do not segregate the wastes and this places a huge challenge on the recyclers. Lack of policies on the implementation of environmental design for plastic products is one of the biggest challenges. Design for recyclability has not been effectively implemented and in most developing economies, it is a huge challenge. A number of waste collection schemes are designed for rigid packaging and this has made flexible packaging problematic when collecting and sorting. Further, the study highlighted that low-weight films and plastic bags have made recycling uneconomical to invest in appropriate sorting and collection facilities. Lack of rationalization on the diversity of plastic materials in terms of subsets of the current usage has placed post-consumer packaging recycling less effective. For example, the separation of rigid plastic containers such as PET, HDPE and PP from co-mingled recyclables would make cross-contamination minimal.

### 3.1.5.3  Current Trends in Plastic Recycling in Developing Economies

A number of staggering problems worldwide relating to plastic manufactured products are unfolding. These problems are in developing as well as developed economies. For example, the United Nations calls plastic pollution in the oceans "planetary crisis" (Waste360, 2018). It is estimated that almost one-third of manufactured plastics end up in the oceans (or soils) and as litter it is not recorded in waste collection data systems. For example, in developing economies, less than 45% of waste is collected for disposal [15].

The trends in managing PSWs in developing economies are shifting. Many studies show that majority of PSWs recovery is performed by the IWCs [9, 11, 14, 24, 35]. The IWCs recover most of the recyclable wastes in developing economies. A study by [11] concludes that the factors that compel the IWCs to recover PSWs are unforeseen to end and therefore, recycling by IWS is likely to continue. The factors that compel the IWCs to recover PSWs include urbanization, poverty, unskilled labor, massive migration, population growth and lack of affordable services. The study conducted by Scheinberg et al. indicates that 30% of wastes generated in major cities of developing economies are recovered by the IWCs. Further, Linzner and Salhofer [20] indicate that 0.93% of the urban population of China is involved in informal waste recovery and recycling.

These studies and many other studies on PSWs recycling affirm that the majority of waste recovery is performed by the IWCs in developing economies. Further, a majority of studies conclude that integration of the IWCs into formalized structures is needed for sustainable waste recoveries [9, 11, 14, 24, 35].

Waste managers, engineers, recyclers and policy makers in developing economies should acknowledge the significant contributions of the IWCs and integrate them into structured systems. In the next 20 years, plastic production is anticipated to double and likely to quadruple by 2050. Therefore, understanding the different trends of plastic recycling is necessary for the establishment of sustainable systems. As described in previous sections, the majority of the recovery is conducted by the IWCs, so they need to be integrated into structured and sustainable systems. Hopewell et al. [16] note that recent trends in plastic recycling rates are increasing and the benefits of recycling are economically and environmentally sustainable. However, significant technological, social and economic challenges related to PSWs recycling exist and managers should design contextual solutions.

In developed economies, the vision for New Plastics Economy established by Ellen MacArthur Foundation is targeted on environmental and financial benefits of a circular economy. The vision lays a 50–30-20 plan for rethinking the future of plastics. It outlines that 50% should be recycled, 30% redesigned and 20% reused (Waste360, 2018). This vision for the future is not only applicable to developed economies but is aligned with the current trends of plastic recycling in developing economies.

## 3.2  Energy Recovery

In its simplest terms, energy recovery involves wastes burning for the production of different types of energy such as heat, electricity and steam. Waste managers, engineers and other stakeholders should understand that energy recovery is only considered a feasible option if other waste treatment options have failed because of economic constraints [2]. For PSWs, various types are utilized in energy recovery processes. From the fact that plastic materials are manufactured from crude oils, there is evidence that shows high calorific value when subjected to heat and burnt. Table 3.1 depicts the calorific value of different single PSWs compared to MSWs and oil. The table depicts that PSWs have a high heating value, thus a convenient energy source.

Considering the low rates of waste collection in most developing economies [15], the option of recovering energy from PSWs is considerable. It is an option that can be considered by understanding the amount of PSWs generated as well as making a comparative analysis of the economic costs involved in recycling PSWs and energy. Environmental concerns are also important to analyze when considering energy

**Table 3.1** Comparison of the calorific value of major plastic and common fuels (*Source* [41] and [22])

| Item | Calorific value (MJ kg$^{-1}$) |
| --- | --- |
| Polyethylene | 43.3–46.5 |
| Polypropylene | 46.50 |
| Polystyrene | 41.90 |
| Kerosene | 46.50 |
| Gas oil | 45.20 |
| Heavy oil | 42.50 |
| Petroleum | 42.3 |
| Household PSW mixture | 31.8 |

recovery options. However, these options will be considered in the other sections later.

Waste managers, engineers and other stakeholders should understand that different types of energy recovery processes exist. These consist of grate technology, and rotary and cement kiln combustion. Outlining the concepts of incineration is necessary as we discuss the option of energy recovery from PSWs.

### 3.2.1 Understanding Incineration of Plastics

Incineration is one of the thermal processing methods for combusting SWs fractions. The primary objective of incineration is to reduce volumes of SWs for sustainable disposal and recover energy released during combustion. Al-Salem et al. [2] note that incineration of PSWs contributes to the reduction in volume of 90–99%, thus reducing reliability on landfilling. In developing economies, advanced investment in technology continues to prevent the successful implementation of incineration. However, with the increase in the amount of PSWs of high calorific values, this is likely to change. It is also necessary to understand that high costs of processing and maintaining operating conditions in energy recovery processes are required.

In most incineration processes, one of the most essential parameters is temperature and this reduces carbon dioxide and nitrogen oxide, as well as increasing nitrogen.

#### 3.2.1.1 Grate Technology

Ordinary MSWs, including PSWs, are usually accepted by MSW incinerators. For PSWs incineration, waste managers should understand that some elements need to be considered. For example, for producing reusable slags, inputting of heavy metals into the incinerator should be minimized. Further, direct incineration is applicable to different types of wastes. In developed economies such as the USA and Germany, a number of incinerators have been designed with a capacity exceeding 10.7 million tons/annual [30]. Advanced technologies in these developed economies have enabled the recovery of PSWs-driven energy value and reducing the consumption of fossil fuel. Many experiments performed by Alliance of Polyurethane (PU) Industry showed that the addition of flexible PU and different fractions of PSWs to MSW generated fuel of high calorific value while the amount of ash generated remained constant. Further, incineration of MSWs with a high content of PSWs is supported in Europe and many countries such as the UK, Sweden, Denmark and Germany supply up to 10% of electrical energy demanded to the local communities.

### 3.2.1.2   Fluidized Bed and Two-Stage Incineration

Weigand et al. [40] describe in detail the concept of bubbling fluidized beds (BFBs) by combusting PSWs and MSWs in high fractions. The study reveals that the incorporation of PSWs in the combustion of MSWs results in 17.6 MJ/kg without increasing the amount of pollutants emissions. The study shows that PSWs are useful for energy recovery, and waste managers should design recovery options for non-recyclable PSWs for utilization in energy recovery processes. PSWs such as PU foam, PE and PS are useful for energy recovery. These PSWs can be incorporated in other wastes used in energy recovery processes such as scrap metals. A study by [34] illustrated the process of BFBs by utilizing a two-stage process to successfully optimize combustion conditions by combusting PU foams from car seats of an automobile. The process resulted in minimal carbon monoxide (CO) and nitroxide (NO) emissions.

### 3.2.1.3   Rotary and Cement Kiln Combustion

Gaseous, solid and fluid waste streams can be processed into energy and useful feedstock using the rotary kiln process. To facilitate high temperatures during combustion, liquid or natural gas carriers can be added. From the fact that the process of a rotary kiln is capable of emitting high temperatures with the addition of other carriers, it is necessary that the technology used in incineration via rotary kilns is understood. BSL technology is the most common technology used and the particle size ranges between $10 \times 10 \times 10$ cm while shredders are utilized for large parts. Rotary kilns are used as incinerators in other industrial schemes and there are known financial benefits for utilizing wastes as fuel [2].

Rotary and cement kiln combustion is conducted using two dominant processes, i.e. wet or dry process. For the wet process, injection of materials is carried out in slurry foam while for the dry process, dry raw materials are injected. The source of the kiln's raw materials determines the type of process that is used and this should be known by the operators. However, the wet process has the disadvantage of demanding more energy compared to the dry process (5000 MJ/ton compared to 3600 MJ/ton clinker).

The utilization of waste as a form of raw materials in rotary and cement kiln processes is an opportunity for waste managers and engineers to develop systems that will enable sustainable recoveries and utilization of these technologies. Supply chain networks for PSWs need to incorporate energy recovery options rather than recycling options alone in developing economies. For example, recovery options from the dumpsites should be set for sustainable recovery of PSWs and other wastes for energy recovery options.

## 3.2.2 *Economic and Environmental Impact of Incineration*

For a number of years, incineration has been a subject of debate in the political, environmental and social arenas. This section discusses the economic and environmental impacts of incineration.

Several environmental concerns are attributed to PSWs incineration. Particular concerns include air pollution (i.e. $CO_2$, $NO_2$ and $SO_2$). Also, a number of volatile organic compounds (VOCs) such as polycyclic aromatic hydrocarbons (PAHs), smoke particulate-bound heavy metals, polychlorinated dibenzofurans (PCDFs) and dioxins are emitted through PSWs incineration [2]. Yassin et al. [43] indicate that ammonia addition to the combustion chamber, flue gas cooling, activated carbon addition, acid neutralization and filtration can be utilized for removing and capturing flue gases in combustion processes.

In the political, social and environmental circles, incineration-based technologies have been a subject of debate. In the petrochemical industry where it is intensely applied, significant pollutants in the form of metal, oxides of nitrogen, particulate matter and sulfur are emitted with an addition of unknown toxicity. These emissions are a danger to the environment and public health. The impacts on public health are risks to cancer and respiratory-related illnesses. The environment is affected in the form of smog formation, acidification, eutrophication, global warming, and animal and human toxicity.

In municipal waste incinerators, approximately 20–35% by weight and approximately 10% by volume of bottom ash are produced. In the incineration processes, the major sources of calorific values are metals and plastics. For PSWs, the major source of highly toxic pollutants is PVC and energy recoveries should put up measures for combating its effects on the environment and public health. It should be known that there are no safe ways of avoiding the production of these pollutants but sustainable costly filters can be used to trap the ashes for disposal in special landfills. In most energy recovery processes, heat exchangers that operate at high temperatures to maximize dioxin production are used but such projects disperse ash to the environment that later enter our food chain.

### 3.2.2.1 Benefits of Energy Recovery

The implementation of the three "Rs" approach to the reduction and protection of the environment is globally known: reduce, reuse and recycle. However, there is the fourth "R" which focuses on energy recovery. This process converts wastes into alternative energy. In developed economies such as the USA, 86 energy recovery facilities across the states process 97,000 tons of wastes per day. The energy generated is used for powering businesses and homes.

The utilization of plastic wastes addresses many challenges existing in developed and developing economies. Several benefits are derived from the utilization of PSWs for energy recovery and include some of these:

- The utilization of PSWs for energy recovery means less plastic wastes are dumped at the landfills and this means reduced illegal disposal or littering of plastic wastes. It is known that plastic wastes are a nuisance to the environment and utilization of plastic means a relief to the environment and the human eyes. Some studies indicate that energy recovery facilities are capable of reducing 90% of the waste disposed at the landfill.
- Technology utilized in the conversion of PSWs to energy has advanced and it is greener than ever before. It allows modernized energy recovery facilities to convert PSWs with cleaner emissions compared to conventional fuels converted in the majority of industries. According to EPA, today's energy recovery technologies restrict the emission of 33 million metric tons of $CO_2$ annually.
- PSWs have a high energy value compared to the other MSWs constitutes. This means plastics contribute to the value of the energy generated while reducing the amount of ash disposed.

## 3.3  Compostable Plastics

Communities are moving away from using post-consumer plastic products and participating in initiatives that are supporting sustainable resource utilization. Some communities are going the route of zero waste while others are engaging in eco-friendly plastic innovates. One of the options is the utilization of products manufactured from compostable materials.

*"Compostable plastics" are plastic materials that are capable of breaking down into water, biomass and carbon dioxide at the rate of cellulose. These plastics disintegrate and distinguish in the compost and do not leave toxic materials.*

### 3.3.1  Understanding Compostable Plastics

Composting plastics has resulted in a number of advantages to the environment and society. For example, compostable plastics are capable of reducing dependence on fossil fuels and can foster the development of several sustainable products. It is necessary to present the differences among traditional, compostable, renewable and biodegradable plastics. Traditional plastics are manufactured from fossil-based plant resources. Renewable plastics are manufactured from living plants resources. In reference to compostable and biodegradable plastics, where and how a plastic biodegrades matter. Biodegradable plastics decompose through naturally occurring actions of living organisms such as home compost, landfill, native soil, industrial compost etc. Compostable plastics biodegrade within the required industrial compost processes as per commercial resale demand. Further, renewable or fossil-based materials can be used to manufacture compostable and biodegradable plastics.

In the drive for the circular economy, compostable plastics compost together with food waste and this is beneficial to the environment. To enhance the process of composting, plastics are not supposed to be sorted from food waste. To large commercial industries that generate huge amounts of food waste, integration of compostable plastics is necessary. Further, it is cardinal for waste managers to understand the difference between compostable and traditional plastics. Traditional plastics and compostable plastics are different and thus cannot be co-mingled in the same recycling stream. If the plastics are mixed and recycling technologies are used, equipment gets damaged as well as contaminates the entire batch. In other words, compostable plastics are not recyclable with traditional plastics.

Several benefits are cited for composting and one of these benefits is related to the reduced carbon footprint. From a greenhouse gas (GHG) scenario, the environmental impact of composting results in reduced methane which is normally generated when the materials are landfilled. However, without control in the compost system, decaying materials produce methane. The greenhouse production depends on the type of waste composted.

### 3.3.2 Current Trends in Compostable Plastics

There is a need for a transition from fossil fuels to sustainable sources. In line with this perspective, compostable plastics manufactured from sources that demand excessive farming should be avoided. An example of such a plastic is thermoplastic starch-based plastic known as ENSO RENEW. These are manufactured from potato starch left after potato processing. This means that the use of leftover starch does not increase farming impacts. Another beneficial advancement is the utilization of methane produced in diaries and landfills as plastic source material. The utilization of methane in plastic manufacturing is an effective environmental innovation as long as the methane is collected from landfills and dairy farms. This advancement means the source of the plastic material is from waste material. This has the advantage of greenhouse gas reduction. The utilization of algae in plastic production is another prospect undergoing.

### 3.3.3 Economic and Environmental Impact of Plastic Composting

Product development is always accompanied by environmental impacts and compostable plastics cannot be isolated. Compostable plastic possesses environmental impacts. However, the critical element is understanding these types of impacts. From the sustainability aspect, the manufacturing of compostable plastics by using sugarcane, potatoes or corns entails environmental sustainability. Nevertheless, environmental impacts attributed to compostable plastics have been reported.

As a result of the massive chemical processing required to transform plants into plastics, the study conducted at the University of Pittsburg concluded that biopolymers are one of the major prolific polluters during the production stage. Kelly adds that during the cultivation of corn used for manufacturing compostable plastics, more nitrogen fertilizer, insecticides and herbicides are utilized compared to other products. Further, in Brazil, sugarcanes are harvested for the purpose of manufacturing plant-based plastics using ethanol. As a result of harvesting sugarcanes for this purpose, vast lands of rain forests are destroyed and also the emissions generated from burnt sugarcanes prior to harvesting. Epidemiological studies indicate that exposure to these emissions causes respiratory diseases.

The fact that material biodegrades into the soil while plants grow from it and the process continues. It is necessary to evaluate the effectiveness of plastics composting while considering the environmental, economic and consumer impacts. Numerous environmental impacts of composting plastics exist.

Nitric oxide production is higher in composting compared to landfilling and this is attributed to intense microbial activities. Further, nitric oxide is 240 times harmful compared to carbon dioxide in contributing to global warming. In addition, the stability of nitric oxide in the environment enables it to last for a long time, thus contributing to ozone-level destruction (Barton and Atwater 2002). Greenhouse gas impacts are dependent on the kind of composted wastes and carbon dioxide in the byproduct. On comparing emissions from landfilling and composting, studies have shown that net GHG emissions are lower in composting compared to landfilling for food discards. However, the opposite is true for yard trimmings. The reason for the high emission in compost compared to landfills is as a result of rapid food waste degradation in landfills and the slow decomposition of yard waste in compost leaves carbon in the soil.

Consideration of economic factors in the sustainability profile is cardinal. For example, the establishment and maintenance of composting facilities and programs demand capital investment due to processing, collection, post-processing and collection of compost. The facilities and programs only become sustainable if the final product is profitable. In most cases, compost facilities are not profitable as the resale value is less than the overhead costs. Collection of material for composting purposes is expensive as it demands separate trucks that should collect small amounts over long distances. This results in increased fuel costs, more wages and more vehicle maintenance to collect the same quantity of material. Further, implementers of composting systems can overlook critical cost and process factors and this might result in the adoption of unsustainable composting systems that produce unpleasant toxic compost. Also, another economic factor for consideration is the production of non-valuable compost. In other words, the profitability of compost facilities is deemed difficult as a result of higher overhead costs.

## 3.4   Landfilling of Plastics

Landfilling is the oldest way of disposing of SW and has been practiced historically. This practice involves disposing of wastes by burying them under the ground. Globally, this practice continues to be the primary methodology of waste treatment strategy by humans. Plastics landfilling remains a key waste treatment methodology and this is shown in Fig. 3.3. The figure shows that landfilling is the most utilized waste treatment system. Nevertheless, in developed economies, legislations regarding the management of PSWs are formulated toward the reduction of PSWs landfilling. In order to develop sustainable waste treatment strategies for managing PSWs, it is necessary to include landfilling as a majority of the PSWs are landfilled. In comparison to composting, recycling and incineration, identifying sustainability options for landfilling demands understanding the details of landfills and how it operates.

Ideally, landfills are engineered to prevent and control toxic wastes, have a reduced compacted volume, are limited to small areas and are covered with layers of soil. To achieve this, an inspection of waste collection vehicles is regulatory to prevent toxic waste to the landfill and upon clearance, disposition of wastes proceeds. The compactors and bulldozers spread and compact the wastes. At the end of the day, compacted wastes are covered with soil or alternative materials such as compost, plastic film or crushed glass. This process is conducted on a daily basis until the space at the landfill is full. More details on the operation of a landfill exist but are not provided in this section.

The environmental and economic sustainability of disposing of PSWs in landfills is attributed to the manner in which plastics are collected, the types of plastics disposed of, the management and design of landfills. The next section focuses on the effects of landfilling in relation to the types of wastes (Fig. 3.2).

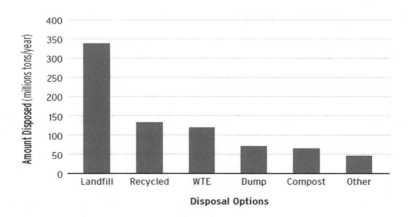

**Fig. 3.2**  Total MSW disposal worldwide [15]

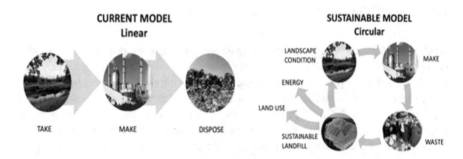

**Fig. 3.3** Current and future model of landfilling (adopted from [38])

### 3.4.1 Understanding the Effects of Landfilling

Dissimilar transformation processes occur in landfills and the majority include the following:

**Chemical Processes**

- Waste materials dragged and solution with leachate;
- Organic compounds dehalogenation and decomposition;
- Adoption of volatile organic compounds (VOC) in the dumping materials;
- Chemical and water compounds evaporation in landfill gases;
- Metal and metallic salts solution redox reactions;
- Waterproof and organic compounds reactions.

**Physical Processes**

- Leachates' migrations and movements;
- Lateral diffusion of landfill gases;
- Movement of dumped materials.

**Biological Processes**

- Gas production process ($CH_4$ and $CO_2$);
- Anaerobic and aerobic digestion.

As a result of the existing transformation processes in landfills, it is necessary and important to discuss the effects of landfills. Nevertheless, the gases generated from the transformation processes can be used for energy production. A number of effects have been revealed to result from landfilling. The effects are categorized into environmental, social and economic problems.

Economic problems of landfilling include initial investments and process investments. Initial investments focus on the main investment required during the establishment of landfill. These include material covering for the bottomland and for the gas as well as leachate control installation. Process investments result from transportation of generated wastes to the landfill and leachate cleaning. Transport effects occur because the environmental and social problems prevent the establishment of landfills near human settlements. Process investment problems are most prevalent in economies where land availability is a challenge.

Environmental problems of landfilling include uncontrolled gas-bags and leaks formation; odor; unrestrained production of insects; greenhouse emissions and atmospheric pollution; and discharge of leachate or pollutants. Danthurebandara et al. affirm that the major environmental impacts are landfill construction, landfill gas and leachate. Thus, to reduce the environmental problems or effects caused by landfills, correct designs are important.

Landfills generate various socioeconomic impacts such as public health caused by landfill gases. The surface water and the ground are contaminated by the landfill leachate. Despite advancements in the design of landfills, emissions from landfills are a major health concern for humans living and working near landfills. In developing economies, a majority of the landfills are not engineered and hence the level of environmental effects to the society is high. It is necessary for waste managers and engineers to understand that exposure to the emissions and contaminants is through direct contact or ingestion of contaminated food and water. A number of studies show that the major source of exposure to harmful substances is caused by the ingestion of contaminated water [18, 39]. Further, these studies affirm that birth weight, child growth, congenital malformations, prematurity and cancers are attributed to landfill emissions. A high risk of developing cancers among people living near landfill sites is revealed in a number of studies [10]. Stomach, intrahepatic bile ducts, bronchus, liver, cervix, prostate and lung are the cancers likely to affect people living near landfills [10]. In addition to the health effects, landfills have an impact on land availability, land value and land degradation. Depending on the distance from the landfills to the houses, several studies affirm that the value of the houses is affected negatively [8, 44]. In other studies, odor, smoke, flies and noise are cited as reasons preventing the public from residing near landfills. Further, demand for enormous landfill space contributes to land scarcity that is needed for human society and ecosystems.

### 3.4.2 Current Trends in Plastic Landfilling

In the hierarchy of WM, the future of landfills is increasing toward recycling and reuse [36]. In Europe, the 1999 directive demands reduced disposals at the landfills. Less than 35% of biodegradable MSW is allowed for disposal at the landfills in most European countries. The strategy is targeted toward reduced waste generation. The strategy implemented is achievable in developing economies. However, implementation is the biggest challenge since most developing economies have SWM problems. Nevertheless, if reduced landfilling is not possible, developing economies should focus on promoting recycling and reusing of SW including PSWs. Energy generation from SW is another alternative developing economies should adopt and as a result of the drive for sustainability, the majority of SW should be used for energy generation.

The WM hierarchy depicts that landfilling is the last option for implementation by waste generators. Indeed, waste should only be landfilled if no alternatives depicted in the WM hierarchy are available. Regardless of the options of WM in the hierarchy, landfilling continues to contribute to the management of SWs. Currently, landfilling has the highest rates globally [15, 29]. This evidence is a call for waste managers and policy makers to acknowledge that not all SWs can be reused, recycled or incinerated. In this regard, landfilling can be the best option for implementation. Further, economic conditions continue to prevent sustainable reusing, recycling and incineration of SWs and PSWs.

In the WM systems, landfills continue to operate as "security networks." Hence sustainable designs that integrate environmental aspects should be considered. In Europe, insulation of landfills with impermeable membranes is a mandatory standard. The membranes have a lifetime of between 50 and 500 years. The purpose of the membranes is to stop all processes within the landfill. In a number of countries, after-care management of landfills is legislated between 30 and 60 years after closure. After-care of landfills is necessary for more than one generation. Sustainable development goals (SDGs) have not isolated landfill management. For example, SDG 6 (clean water and sanitation) is a goal that is compromised by poor management of landfills and hence the need to design sustainable landfills. However, there is no international definition for sustainable landfills. Figure 3.3 shows a schematic diagram of sustainable landfilling [38]. According to Townsend [37], sustainable landfilling highlights the following terms: no environmental threats, completion, stability and end technologies. Stability refers to "functionally stable" landfill that is not a threat to the environment and human health. Completion means the landfill is biologically, physically and chemically stable and without risk to the environment.

Several landfilling concepts have evolved and it is necessary for waste manages, engineers and interested stakeholders to understand. Despite having last place on the waste hierarchy, as defined by the EU waste directive (2008/98/EC), it is still applied as a WM strategy because of the increasing waste generation and the lack of sustainable techniques to treat different types. Nevertheless, landfilling has evolved and one approach is engineered bioreactor landfills. A controlled degradation system is allowed to guarantee sustained stability [33]. According to [33], a bioreactor landfill utilizes enhanced microbiological processes to stabilize and transform decomposable organic waste within 5–8 years. Bioreactor sanitary landfills increase waste convention rates, decomposition and process effectiveness compared to the traditional landfills. Several advantages are presented by bioreactors and these include reduced environmental impact, enhanced landfill gas generation rates, reduced landfilling costs, reduced leachate treatment costs, reduced after-care costs and reduced maintenance costs. Currently, a handful of bioreactor landfills are in operation and this is attributed to fairly new designs. Nevertheless, the economic and environmental advantages presented by the bioreactor will fast-track the conversion of all landfill designs. Furthermore, bioreactors landfills are not landfills but similar to anaerobic digesters in which organic substances biodegrade into methane, air, soil and water. The methane is utilized for energy purposes while non-biodegradable materials are recycled or reused.

The concept of enhanced landfill mining (ELFM) reduces potential hazardous emissions while valorizing resources contained within. According to [31], the amounts of vital materials have accumulated in landfills. Globally, approximately 393 million metric tons of copper are situated in landfill deposits. Apart from old landfills, ELFM is implemented on new landfills while considering temporary storage. In this regard, landfills are future mines for materials that cannot be recycled with current technologies [32]. Kamaruddin et al. [17] alludes to the existence of economic potential from ELFM projects. Several advantages of ELFM have been highlighted and include climate gains, energy and material utilization, land reclamation and job creation. Powell et al. [31] adds that tailored support systems and policy decisions including incentives for nature restoration, material recycling and energy utilization are required.

# References

1. SubsTech, Substances and Technologies, *Link to Plastic Recycling* (2006). Available http://www.substech.com/dokuwiki/doku.php?id=plastics_recycling
2. S.A. Al-Salem, P. Lettieri, J. Baeyens, Recycling and recovery routes of plastic solid waste (PSW): a review. Waste Manage. **29**, 2625–2643 (2009)
3. M.P. Aznar, M.A. Caballero, J.A. Sancho, E. Francs, Plastic waste elimination by co-gasification with coal and biomass in fluidized bed with air in pilot plant. Fuel Process. Technol. **87**(5), 409–420 (2006)
4. C. Barlow, *Intelligent Recycling (Presentation)* (University of Cambridge, Department of Engineering Institute for Manufacturing, 2008)
5. A. Bartl, Ways and entanglements of the waste hierarchy. Waste Manage. **34**, 1–2 (2014)
6. Bio-Intelligence, *Study on an increased mechanical recycling target for plastics*. Final report prepared for Plastic Recyclers Europe (2013)
7. P. Brachet, L.T. Høydal, E.L. Hinrichsen, F. Melum, Modification of mechanical properties of recycled polypropylene from post-consumer containers. Waste Manage. **28**(12), 2456–2464 (2008)
8. Y.C.H. Chen, Evaluating greenhouse gas emissions and energy recovery from municipal and industrial solid waste using waste-to-energy technology. J. Clean. Prod **192**, 262–269 (2018)
9. T.M. Coelho, PET containers in Brazil: opportunities and challenges of a logistics model for post-consumer waste recycling. Resour. Conserv. Recycl. **3**(55), 291–299 (2011)
10. J. Duggan, The potential for landfill leachate treatment using willows in the UK—a critical review. Resour Conserv. Recycl. **45**, 97–113 (2005)
11. C. Ezeah, J.A. Fazakerley, C.L. Roberts, Emerging trends in informal sector recycling in developing and transition countries. Waste Manage. **33**, 2509–2519 (2013)
12. R. Frosch, N. Gallopoulos, Strategies for manufacturing. Sci. Am **261**, 144–152 (1989)
13. C. Giacovelli, *Plastic Waste Management* (International Environmental Technology Centre (IETC), Osaka, Japan, 2017)
14. J. Gutberlet, *Recovering Resources—Recycling Citizenship: Urban Poverty Reduction in Latin America* (Ashgate, Aldershot, 2008)
15. D. Hoornweg, P. BhadaTata, *What a Waste: A Global Review of Solid Waste Management. Urban Development Series Knowledge Papers* (World Bank, Washington, DC, 2012)
16. J. Hopewell, R. Dvorak, E. Kosior, Plastics recycling: challenges and opportunities. Phil. Trans. R. Soc. B **364**, 2115–2126 (2009)
17. M.A. Kamaruddin, M.S. Yusoff, L.M. Rui, A.M. Isa, M.H. Alrozi, R. Zawawi, An overview of municipal solid WM and landfill leachate treatment: Malaysia and Asian perspectives. Environ. Sci. Pollut. Res. **24** (2017)

18. E. Koda, S. Zakowicz, Physical and hydraulic properties of the MSW for water balance of the landfill, in*Proceedings of the 3rd International Congress on Environmental Geotechnics* (Lisbon, Portgual, 1998), pp. 217–222
19. Y. Lei, Q. Wu, F. Yao, Y. Xu, reparation and properties of recycled HDPE/natural fibre composites. Composites Part A: Appl. Sci. Manuf. **38**(7), 1664–1674 (2007)
20. R. Linzner, S. Salhofer, Municipal solid waste recycling and the significance of informalsector in urban China. Waste Manage. Res. **32**(9), 896–907 (2014)
21. E. Manzini, C. Vezzoli, *Odesenvolvimento de produtos sustentáveis* (Edusp, São Paulo, 2005)
22. M.L. Mastellone, *Thermal treatments of plastic wastes by means of fluidized bed reactors.* Ph.D. Thesis,, Naples: Department of Chemical Engineering, Second University of Naples, Italy (1999)
23. W. McDonough, M. Braungart, *Cradle to Cradle: Remaking the Way We Make Things* (North Point Press, New York, 2002)
24. M. Medina, *The World's Scavengers: Salvaging for Sustainable Consumption and Production* (AltaMira Press, 2007)
25. C. Meran, O. Ozturk, M. Yuksel, Examination of the possibility of recycling and utilizing recycled polyethylene and polypropylene. Mater. Des. **29**(3), 701–705 (2008)
26. B.G. Mwanza, *An African Reverse Logistics Model for Plastic Solid Wastes* (University of Johannesburg, PhD Report, Johannesburg, 2018)
27. Opportunity, Sea of, *Supply Chain Investment Opportunities to Address Marine Plastic Pollution* (2017)
28. E.B.A.V. Pachecoa, L.M. Ronchettia, E. Masanetb, An overview of plastic recycling in Rio de Janeiro. Resour. Conserv. Recycl. 140–146 (2012)
29. N. Pietzsch, J.D.L. Medeiros, J.F. Ribeiro, Benefits, challenges and critical factors of success for zero waste: A systematic literature review. Waste Manag. 324–353 (2017)
30. Pollution Issue, Incineration, Ho:Li. Available at: pollutionissues.com (2007)
31. J.T. Powell, J.C. Chertow, M. Pons, Waste informatics: Establishing characteristics of contemporary U.S. landfill quantities and practices. Environ. Sci. Technol. **50**, 10877–10884 (2016)
32. G.E. Raúl, S. Morales, R. Toro, A. Luis, M.A. Leiva, G. Morales, Landfill fire and airborne aerosols in a large city: Lessons learned and future needs Air Qual. Atmos. Health **11**, 111–121 (2018)
33. S. Renou, J.G. Givaudan, S. Poulain, F. Moulin, P. Dirassouyan, Landfill leachate treatment: Review and opportunity. J. Hazard. Mater. **150**, 468–493 (2008)
34. Y. Rogaume, F. Jabouille, M. Auzanneau, J.C. Goudeau, *Proceedings of the 5th International Conference on Technologies and Combustion for a Clean Environment* (Lisbon, Portugal, 1999) pp. 345–351
35. A. Scheinberg, *Value Added, Modes of Sustainable Recycling in the Modernisation of Waste Management Systems.* Ph.D. dissertation. Wageningen University, The Netherlands, The Netherlands: WASTE., Gouda (2011)
36. K. Stoeva, S.R. Alriksson, Influence of recycling programmes on waste separation behavior. Waste Manage. **68**, 732–741 (2017)
37. T.G. Townsend, J. Powell, P. Jain, Q. Xu, T. Reinhart, D Tolaymat, Planning for sustainable landfilling practices. sustainable practices for landfill design and operation, in*Sustainable Practices for Landfill Design and Operation* (Springer, New York, 2015)
38. M.D. Vaverková, Landfill impacts on the environment—review. Geosciences **2019**(9), 431 (2019)
39. S. Wang, Y. Shen, Performance of an anaerobic baffled reactor (ABR) as a hydrolysis-acidogenesis unit in treating landfill leachate mixed with municipal sewage. Water Sci. Technol. 115–121 (2000)
40. E. Weigand, J. Wagner, G. Waltenberger, Energy recovery from polyurethanes in industrial power plants. Abfall J. **3**, 40–45 (1996)
41. E.A. Williams, P.T. Williams, The pyrolysis of individual plastics and plastic mixture in a fixed bed reactor. J. Chem. Technol. Biotechnol. **70**(1), 9–20 (1997)

42. D.C. Wilson, C. Velis, C. Cheeseman, Role of informal sector recycling in waste management in the developing countries. Habitat Int. **30**, 787–808 (2006)
43. L. Yassin, P. Lettierim, S.J.R. Simons, A. Germana, Energy recovery from thermal processing of waste: a review. Eng. Sustain. (Proc ICE) **158**(ES2), 97–103 (2005)
44. G. Zappini, P. Rossi, D. Cocca, Performance analysis of energy recovery in an Italian municipal solid waste landfill. Energy **35**, 5063–5069 (2010)
45. H. Zhang, Z.G. Wen, Consumption and recycling collection system of PET bottles: a case study of Beijing, China. Waste Manage. **34**, 987–998 (2014)
46. K.M. Zia, H.N. Bhatti, I.A. Bhatti, Methods for polyurethane and polyurethane composites, recycling and recovery: a review. React. Funct. Polym. **67**(8), 675–692 (2007)

# Chapter 4
# Sustainability in PSWs Management Technologies: A Review

The evolution of sustainability is discussed for the purposes of presenting the current status in developing economies. From the definitions of sustainability presented by the United Nations, the developing economies are working with the same definition. In this regard, the chapter presents an overview of sustainability by discussing the three pillars of sustainability. Indicators of sustainability in PSWs management technologies are presented with the intent of influencing the decisions waste managers, engineers and stakeholders make. To complement the discussions on sustainability in PSWs management technologies, a life cycle assessment of PSWs is discussed by considering sustainable ways of achieving sustainability in PSWs management technologies for developing economies.

## 4.1 What is Sustainability in Developing Economies?

Universally, sustainability has no agreed definition as there are many different views on how it is achieved. The concept of sustainability was first published in the Brundtland report in 1987. In 1992, the definition of sustainability stemmed from the sustainable development concept. The emphasis on sustainability was to warn the United Nations on the environmental demerits of globalization and economic development. Today, sustainability focuses on the present and future generation needs. The following are some of the definitions of sustainability:

*"Sustainability" is an investigation of natural systems and their ability to produce, function and remain diverse for the purpose of sustaining an ecological balance* [14].

*"Sustainability" is the ability for the present generation to meet their needs without endangering the abilities of the future generation to meet their needs* [1].

© The Author(s), under exclusive license to Springer Nature Switzerland AG, part of Springer Nature 2022
B. G. Mwanza and C. Mbohwa, *Sustainable Technologies and Drivers for Managing Plastic Solid Waste in Developing Economies*, SpringerBriefs in Applied Sciences and Technology, https://doi.org/10.1007/978-3-030-88644-8_4

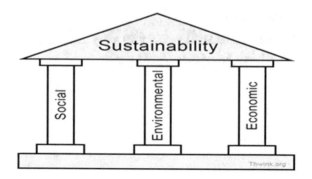

*"Sustainability" is sustained structures for meeting present needs without compromising the abilities of the future generations to meet their own needs* [7].

The definitions of sustainability focus their attention on the ability of the present and future generations to attain their needs. It promotes satisfactory levels of achievement in the quality of health, life and education for cultures; seeks cohesion among communities and promotes societal development. Waste managers and engineers should understand that a number of human challenges, such as WM, climate change etc., can only be solved from a global perspective and through the promotion of sustainable development. In essence, sustainable programs stem from an institution's commitment to the pillars of sustainability or the "triple bottom line." It is applicable from an individual and institutional perspective and it is mostly a balancing act. Figure 4.1 depicts the sustainability pillars.

Sustainability pillars provide a potent tool for delineating the whole sustainability problem. These pillars should be in a balancing act because weakness in any one of the pillars means the whole system is unsustainable. To address PSWs management from a sustainability perspective, waste experts should not isolate any of the pillars. Nevertheless, most international or national efforts to solve sustainability-based problems have focused on a single pillar at a time. Examples are the United Nations Environmental Programme (UNEP), environmental NGOs and the Environmental Protection Agencies (EPA). These have focused attention on the environmental pillar. In the case of the World Trade Organization (WTO) and the Organization for Economic Cooperation and Development (OECD), the focus has been on economic growth. In the case of the United Nations, it attempts to focus on all the pillars but some decisions prevent the balancing act and strife attention to the economic pillar. In developing economies, the three pillars have not been balanced when it comes to achieving sustainability. For example, the economic pillar has been given more attention in developing economies by the United Nations. In order for sustainability to be understood by the waste experts, the three pillars are discussed in detail.

Economic sustainability focuses attention on supporting a defined level of economic production for unspecified periods of time. From the period of the great recession in 2008, attention has been given to economic sustainability and progress on social and environmental sustainability has been hindered. Globally, the problem of PSWs management has been their focus on the economic aspects but the environmental and social impacts of successfully managing PSWs have not been given much attention. For example, in developing countries, PSWs management is a challenge as the local authorities focus more on the economic roadblocks instead of paying attention to the social and environmental aspects that can drive sustainable PSWs management.

Social sustainability focuses its attention on the well-being and harmony of its social system. For example, problems such as endemic poverty, WM and others are addressed to provide social sustainability. For waste experts, attention should focus on the social challenges preventing sustainable PSWs management. In a number of studies, the social aspect of involving society has been presented. Key socioeconomic factors have been found to impact the sustainable management of waste [27]. The findings of these studies provide evidence that the involvement of society in waste-related problems is required in order to achieve sustainability.

Environmental sustainability focuses on the environment's capability to support a certain level of natural resource extraction rates and environmental quality indefinitely. Globally, low attention is given to the environmental pillar as a result of the attention given to the economic pillar. The environmental pillar continues to receive low priority because of the economic decisions implemented and this further affects the society directly through health-related outbreaks. Figure 4.2 depicts the relationship between the three pillars of sustainability.

According to Thwink.org (2018), there is no international organization that is resolving the sustainability problem by way of using all three pillars. Figure 4.2 shows that there is a bigger relationship among the three pillars. In order for waste experts to understand sustainability and resolve the challenges of WM facing developing economies, it is necessary for the experts to picture the globe as a set of interconnected systems. Because of the fact that sustainability is concerned with the present and future generation needs, the diagram shows that the biosphere is the largest system and it accommodates the human system. In the human system, the social and economic pillars of sustainability exist. In the diagram, the social contract is formed by the people in order to improve their welfare. For WM, the

**Fig. 4.2** The bigger picture of sustainability pillars (Thwink.org [37])

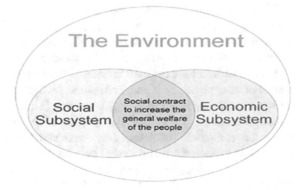

social contract is the constitution of stakeholders that are driven to solve the challenges facing the WM sector. Waste experts should identify how social organizations
and stakeholders can contribute to resolving the sustainability problem from a social
perspective. The social contract binds the economic and social systems of groups
of individuals. This implies the people (social subsystem) work to maximize the
economic system's output. It is important for waste experts to see the overall system
in the manner depicted in Fig. 4.2 as it indicates that environmental sustainability has
the highest rank, indicating that the lower the environmental impacts are, the lower
the social system will be impacted, and hence the lower the output expected from
the economic systems.

## 4.2    Indicators of Sustainability in PSWs Management Technologies

Global efforts for compelling the reorientation of PSWs management systems toward
sustainability are ongoing, and developing economies should be deeply engaged in
this transition. Regardless of how the level of attention given to sustainability changes
from nation to nation and is connected to economic standing, the sustainability of
any WM practice relies on cost-effectiveness abilities and is mostly defined by the
economic standing of a society. In developing economies, considerate efforts are
being taken to maximize the recycling rates in order to decrease the pressure on
landfills. This is also seen from a global dimension as the world retrieves sustainability and measures to decrease material consumption. Based on the 3R initiatives
that were formulated with the concept of "reduce, reuse and recycle" of waste materials with the intent of decreasing the final amount of waste to landfill sites (MoE
Japan [38]). A system for SSWM should be designed within the society's capacity
to cover different issues such as technological availability or financial affordability
as combinative capacities of an adjoining environment. The system should be pragmatic in design and have the capacity to provide an approach that is realistic to
improvements.

The application of indicators as assessment tools for measuring the sustainability
of PSWs management technologies is relevant as it provides information cardinal
for policy makers, waste managers and engineers to use in decision-making. Due to
the importance of the PSWs management problems, it is completely important that
necessary indicators are presented and discussed.

Several authors have defined "indicators" in different ways and this has presented
a number of contradictions and ambiguities on the general concept of an indicator.
To some authors, an indicator means variable [8]; a plan [19]; a parameter [22]; a
statistical measure (Tunstall [41]); a portion of data [6] etc. The above definitions
reveal the complexity of defining an "indicator." However, as stated by [12] an
indicator is an element that provides data on difficult elements and is applied as a
benchmarker for decision-making. In technologies for managing PSWs, the use of
indicators is important, but focusing on the definition of an indicator alone will leave
a gap. Therefore, it is necessary that the definition for sustainability indicators is
presented in this book.

*According to Zabaleta (2008), a sustainability indicator provides information on the sustainability of a system, activity and/or process.* It provides information that guides the decision-making process, and in this case, the aspects of economic, social and environmental attributes are considered. The merit of utilizing indicators gives the possibility of identifying problems, developing policies and easing the complexity of a system. This is important for solving sustainability issues because the cause and effect chains between the environmental, social and economic aspects are complex.

## 4.3  Life Cycle Assessment of Post-consumer PSWs

Life cycle assessment (LCA) was established a few decades ago as a critical tool for assessing environmental problems associated with material cycle and chemical processes [4]. It is a management approach and environmental accounting tool that utilizes all aspects of environmental releases and resource use connected with an industrial system from creation to disposal. LCA takes a holistic environmental interaction that focuses on various actions from the extraction of natural materials, manufacturing and distribution of energy, through utilization, reuse and final disposal of the product. It is a tool that was designed to compare but not evaluate, and hence helping decision makers to compare major environmental impacts when implementing the alternative course of actions.

Over the past decades, LCA has been used as a critical tool for comparing environmental burdens brought about by a number of EoL products. PSWs are a major EoL waste that continues to increase and pose environmental problems. Despite the fact that PSWs can be recovered and utilized, landfills continue to fill up, and this shows that landfilling is still a considered waste treatment option. For example, in developing economies, PSWs are a major problem, as discussed in the previous chapter.

The common routes for disposing of PSWs are landfilling, energy recovery and mechanical recycling (Lazarevic et al. [39]). It is necessary to indicate that majority of PSWs or MSWs end up in landfills or waste to energy facilities as a result of the non-compatibility of the materials with the recycling and reuse technologies [18]. In this regard, a number of LCA studies on plastics have been conducted. In China, Gu et al. [16] assessed the LCA of mechanical plastic recycling. The study revealed that mechanical recycling of PSWs is environmentally sustainable compared to virgin plastic manufacturing. The impact of virgin plastic production on the environment is four times more than mechanical recycling of PSWs. Coupled with the emissions released during mechanical recycling of PSWs, the approach is fairly environmentally friendly. In Singapore, [17] conducted a study on PSWs recovery through an LCA approach. The recovery of PSWs was performed for the purposes of recycling, energy and fuels generation. Pyrolysis, mechanical recycling and gasification were used as waste treatment options. The results showed that LCA outcomes are based on the weighting factors and the system boundaries considered during analysis. Demetrious and Crossin [40] compared and examined waste treatment through gasification-pyrolysis, landfilling and incineration. The study showed

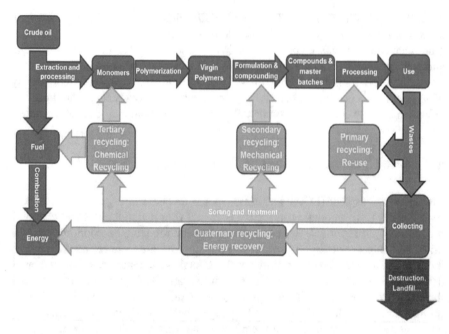

**Fig. 4.3**   The typical life of plastics [9]

the relevance of pyrolysis on GHG reduction potential and indicated that landfilling needs less energy and is much preferred compared to gasification–pyrolysis. Other studies have also looked at the LCA of PSWs [3, 13, 15, 23]. From the evaluation of these studies, it can be concluded that, globally, PSWs are still a major concern because of the environmental impacts. Despite the implementation of pre-treatment and recycling approaches that are beneficial for impacts reduction, less plastics are recycled and hence there is a need to emphasize the significance of post-treatment. In this case, LCA as a tool designed for evaluating environmental impacts can be used for analyzing the burdens that can be avoided from the processes of WM (i.e. recycling etc.). Published researches on LCA of PSWs indicate that thermochemical post-treatments (i.e. pyrolysis or gasification, incineration) have decreased environmental impacts compared to landfilling. Figure 4.3 shows the different stages in the life cycle of post-consumer plastic product. The stages depicted in the figure are usually assessed during the LCA analysis of PSWs.

## 4.4   Way Forward to Sustainability in PSWs Management

The momentum targeted toward beating PSWs pollution continues to grow in recent years. Globally, several policies and efforts for limiting plastic pollution continue to receive limited success. The case for developing economies on the matter is still in its infancy. Nevertheless, several studies have looked at addressing PSWs from the sustainability point of view. In this book, the following are recommended:

- Regional PSWs management treaties should be established for tackling plastic pollution from the point of origin. Consideration of PSWs management treaties presents an opportunity for supporting economies within regional boundaries by enhancing waste collection and recycling systems. Further collaboration with other regional communities such as the African Union should be encouraged. The regional communities should establish directives purposed for ending plastic pollution. An example of such a directive that is aimed at PSWs management is the EU directive.
- Awareness of the environmental merits of PSWs recycling should be provided to all stakeholders. The exercise should be intensified in learning institutions (kindergarten, primary, secondary, tertiary, colleges, universities etc.), companies and organizations. Social media platforms such as radio, television, internet etc. should be used. This recommendation is supported in the literature by studies conducted by [28] and [2].
- The importance of IWCs in PSWs management should be effectively communicated to the communities. In developing economies, the IWCs play a significant in the recycling systems of PSWs, and hence the public should be aware of this key driver. To enhance and intensify value addition in the supply chain of PSWs, integration of the IWCs should be implemented. Gunsilius [33]; Atienza [5] and Medina [20] support this recommendation.
- The majority of developed economies have legalized PSWs recovery and recycling and this has contributed to sustainable management. Therefore, this is recommended for developing economies. Nation-wide PSWs recycling legislations should be mandated by policy makers. Sidique [25] highlight that legislations and regulations contribute to the sustainable management of PSWs.
- EPR policies contribute to the sustainable implementation of RLs systems for PSWs. In this regard, the governments of developing economies should mandatorily introduce EPR in the plastic manufacturing and distributing companies. Further, the plastic industry should work with the WM companies (public and private) to ensure that the physical and financial aspects of EPR are met. Xevengos et al. [28] and Crux et al. [10] support this recommendation.
- Source segregation is key in achieving sustainability in PSWs management. Separation of PSWs from other wastes at the point of generation contributes to an efficient sorting process. Therefore, to promote this recommendation, waste receptacles for recyclable wastes (PSWs) should be provided. Matter et al. [36]; Chiang et al. [35] support this recommendation.
- Effective waste collection persists as a major challenge in developing economies. Therefore, integrated collection systems that involve the IWCs should be established and formalized to improve the recovery of PSWs and other wastes. Zhang and Wen [29] and Zhang et al. [24] support this recommendation.
- Consideration of the effect of socioeconomic factors (gender, income, age and education) in the implementation of successful PSWs recycling networks is important. The study conducted by [21] showed the impact of socioeconomic factors on participation. Other studies have shown the effect of socioeconomic factors [11, 26].

- Establishment of quality standards in the PSWs supply chain management for the purpose of improving the recoveries and price systems. BIO-Intelligence [34] affirms that improved quality standards lead to improved sustainable price systems.
- Increment and establishment of recycling facilities is key to improving recovery rates. This recommendation can work if the above-suggested recommendations are considered. For example, EPR, awareness and source segregation are but a few of the recommendations that can lead to the establishment of more recycling facilities. Rispo et al. [31] affirm that infrastructure contributes to resident participation in WM activities.
- Structured pricing establishments are cardinal to both the informal and formal sectors involved in PSWs recovery and recycling. Currently, inconsistencies exist in the pricing systems of recovered PSWs. Gutberlet [32] notes that extreme fluctuation prevails in the IWS and proper price systems should be established.
- As discussed in the previous sections of this book, a number of waste treatment options exist. To improve sustainability in the plastic industry and management, the costs of incineration, landfilling and recycling should be comparable. This will enable companies to implement sustainable systems.

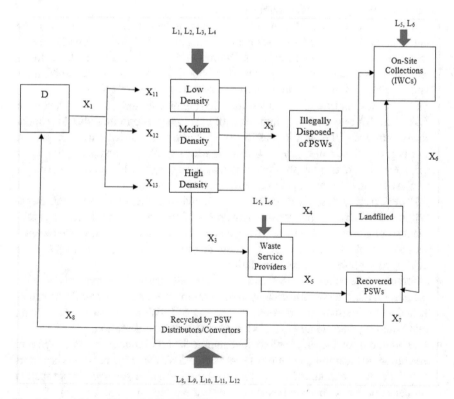

**Fig. 4.4** An African RL model for PSWs [21]

- Incentives provision to the communities is key for effecting participation behavior in recycling programs for PSWs. Welfens et al. [30] affirm that incentives are critical for initiating sustainable behavior styles and his study supports incentive provision to households by manufacturers.
- Adoption of structured RL systems for PSWs recycling. A study by [21] proposed an African RL system for recovering and recycling PSWs. Figure 4.4 depicts the proposed system.

# References

1. Growing responsibility, *World Commission on Environment and Development (WCED)*. Our Common Future (Oxford University Press, Oxford, 2018)
2. R. Afroz, A. Rahman, M.M. Masud, R. Akhtar, The knowledge, awareness, attitude and motivational analysis of plastic waste and household perspective in Malaysia. Environ. Sci. Pollut. Res. **23**, 2304–2315 (2017)
3. S. Al-Salem, S. Evangelisti, P. Lettieri, Life cycle assessment of alternative technologies for municipal solid waste and plastic solid waste management in the Greater London area. Chem. Eng. J. **244**, 391–402 (2014)
4. A. Antelava, S. Damilos, S. Hafeez, G. Manos, S.M. Al-Salem, B.K. Sharma, K. Kohli, A. Constantinou, Plastic solid waste (PSW) in the context of life cycle assessment (LCA) and sustainable management. Environ. Manage. **64**(2), 230–244 (2019)
5. V. Atienza, Sound strategies to improve the condition of the informal sector in waste management, in *ERIA Research Project Report, 2009, Chiba, Japan: Institute of Developing Economies* (Japan External Trade Organization, 2010)
6. J.A. Bakkes, G.J. van den Born, J.C. Helder, R.J. Swart, C.W. Hope, J.D.E. Parker, *An Overview of Environmental Indicators: State of the Art and Perspectives UNEP/EART* (UNEP, Environmental Assessment Sub-Programme, Nairobi, 1994)
7. Bruntland, *Report for the World Commission on Environment and Development* (1992)
8. S. Chevalier, R. Choiniere, L. Bernier, et al., User Guide to 40 Community Health. *User Guide to 40 Community Health Indicators* (Community Health Division, Ottawa: Health and Welfare Canada, 1992)
9. R. Clift, Overview clean technology—the idea and the practice. J. Chem. Technol. Biotechnol. **68**(4), 347–350 (1997)
10. N.F. Cruz, S. Ferreira, M. Cabral, P. Simões, R.C. Marques, Packaging waste recycling in Europe: is the industry paying for it? Waste Manage. 298–308 (2014)
11. T. Domina, K. Koch, Convenience and frequency of recycling: implications for including textiles in curbside recycling programs. Environ. Behav. **34**, 216–238 (2002)
12. R. Gras, M. Benoit, J.P. Deffontaines, M. Duru, M. Lafarge, A. Langlet, P.L. Osty, *Le Fait Technique en Agronomie, Activité Agricole, Concepts et Méthodes d'Étude* (L'Hamarttan, Institut National de la Recherche Agronomique, Paris, France, 1989)
13. M. Gunamantha, Life cycle assessment of municipal solid waste treatment to energy options: case study of KARTAMANTUL region, Yogyakarta. Renew. Energ. **41**, 277–284 (2012)
14. http://www.epa.gov/sustainability/basicinfo.htm. n.d.
15. D. Iribarren, J. Dufour, D.P. Serrano, Preliminary assessment of plastic waste valorization via sequential pyrolysis and catalytic reforming. J. Mater. Cycles Waste **14**(4), 301 (2012)
16. J. Gu, H. Xu, C. Wu, Thermal and crystallization properties of HDPE and HDPE/PP blends modified with DCP. Adv. Polym. Technol. **33**(1) (2014)
17. H.H. Khoo, LCA of plastic waste recovery into recycled materials energy and fuels in Singapore. Resour. Conserv. Rec. **145**, 67–77 (2019)
18. M. Margallo, R. Aldaco, A. Bala et al., Contribution to closing the loop on waste materials: valorization of bottom ash from wastetoenergy plants under a life cycle approach. J. Mater. Cycles Waste **20**(3), 1507–1515 (2018)

19. D. McQueen, H. Noak, Health promotion indicators: current status, issues and problems. Health Promotion **3**, 117–125 (1988)
20. M. Medina, *Globalization, Development, and Municipal Solid Waste Management in Third World Cities* (2002). Accessed 12 Aug 2016. Available at: http://depot.gdnet.org/cms/confer ence/papers/5th_pl5.2_martin_medina_martinez_paper.pdf.
21. B.G. Mwanza, *An African Reverse Logistics Model for Plastic Solid Wastes* (University of Johannesburg, PhD Report, Johannesburg, 2018)
22. OECD, *Organization for Economic Cooperation and Development Core Set of Indicators for Environmental Performance Reviews* (OECD, A Synthesis Report by Group on the State of the Environment, Paris, 1993)
23. L. Rigamonti, M. Grosso, M. Giugliano, Life cycle assessment for optimising the level of separated collection in integrated MSW management systems. Waste Manag. **292**, 934–944 (2009)
24. S. Rodrigues, G. Martinho, A. Pires, Waste collection systems. Part A: a taxonomy. J. Clean. Prod. **113**, 374–387 (2016)
25. S.F. Sidique, F. Lupi, S.V. Joshi, The effects of behaviour and attitudes on drop-off recycling activities. Resour. Conserv. Recycl. **54**, 163–170 (2010)
26. A.M. Troschinetz, J.R. Mihelcic, Sustainable recycling of municipal solid waste in developing countries. Waste Manage. **29**(2), 915–923 (2009)
27. P. Vicente, E. Reis, Factors influencing households' participation in recycling. Waste Manage. Res. **26**, 140–146 (2008)
28. D. Xevgenos, C. Papadaskalopoulou, V. Panaretou, K. Moustakas, D. Malam, Success stories for recycling of MSW at municipal level. Waste Biomass. Valor. **6**, 657–684 (2015)
29. H. Zhang, Z.G. Wen, Consumption and recycling collection system of PET bottles: a case study of Beijing, China. Waste Manage. **34**, 987–998 (2014)
30. M.L. Welfens, J. Nordmann, A. Seibt, Drivers and barriers to return and recycling of mobile phones. Case studies of communication and collection campaigns. J. Cleaner Prod., 1-14 (2015)
31. A.I.D. Rispo, P.J. Williams, P.J. Shaw, Source segregation and food waste prevention activities in high density households in a deprived urban area. Waste Management **44**, 15-17 (2015)
32. J. Gutberlet, *Recovering Resources—Recycling Citizenship: Urban Poverty Reduction in Latin America* (Ashgate, Aldershot, 2008)
33. E. Gunsilius, S. Spies, S. García-Cortés, M. Medina, S. Dias, A. Ruiz, Recovering resources, creating opportunities-Integrating the informal sector into solid waste management. The Deutsche Gesellschaft für Internationale Zusammenarbeit (GIZ) (2012)
34. BIO Intelligence Service, Study on an increased mechanical recycling target for plastics. final report prepared for Plastic Recyclers Europe (2013)
35. T.A. Chiang, Z.H. Che, Z. Cui, Designing a multistage supply chain in cross-stage reverse logistics environments: application ofparticle swarm optimization algorithms. Sci. World J. (2013)
36. A. Matter, M. Dietschi, C. Zurbrügg, Improving the informal recycling sector through segregation of waste in the household–The case of Dhaka Bangladesh. Habitat International, 150-156 (2013)
37. Thwink.Org, 2018. Finding and resolving the root cause of the sustainability problem. https://www.thwink.org/sustain/glossary/ThreePillarsOfSustainability.htm Accessed 24 March 2020 (2018)
38. Ministry of the Environment, Annual Report on the Environment in Japan 2006. https://www.env.go.jp/en/wpaper/2006/index.html Accessed 24 March 2020 (2006)
39. D. Lazarevic, E. Aoustin, N. Buclet, N. Brandt, Plastic waste management in the context of a European recycling society: Comparing results and uncertainties in a life cycle perspective. Resource. Conserv. Recycling **55**(2), 246-259 (2010)
40. A. Demetrious, E. Crossin, Life cycle assessment of paper and plastic packaging waste in landfill, incineration, and gasification-pyrolysis. J. Mater. Cycles Waste Manag. **21**(4), 850-860 (2019)
41. D. Tunstall, The growing importance of scientific rules of thumb in developing indicators of resource sustainability. In Prepared for the International Conference on Earth Rights and Responsibilities (1992)

# Chapter 5
# Drivers to Sustainable PSWs Management: A Review

To achieve sustainable management of PSWs, numerous drivers are key for influencing implementation. The chapter discusses in detail the numerous drivers that influence sustainable management of PSWs from the developed and developing economy perspectives. The drivers include incorporation of IWS into formalized systems, economic incentives, appropriate technology considerations, societal participation and awareness recycling schemes, regulations and legislations, collection and segregation systems, household education, institutional arrangement, training the informal waste sector and local recycled material markets. Applicability of the drivers to the developing economy context is emphasized by understanding the drivers and studying how each driver was achieved from a developed economy perspective. In this regard, feasibility studies are recommended for determining the relevance of each driver in each context.

## 5.1 Incorporation of IWS into Formalized Systems

Recovery and recycling of wastes is a livelihood for most marginalized societies in developing countries. Unlike developed economies, i.e. the USA, recovery and recycling of wastes is an informal activity [3]. According to Gunsilius et al. [19], the IWS provides important services for the communities and enhances the hygienic conditions of the environments. For this and many other benefits that the IWS contributes to society, many studies recommend integration of the IWS into formalized systems as a strategy to SSWM. For example, small to medium enterprises use recovered materials as raw materials in their production processes and EoL plastic packaging products are typical examples.

B. G. Mwanza and C. Mbohwa, *Sustainable Technologies and Drivers for Managing Plastic Solid Waste in Developing Economies*, SpringerBriefs in Applied Sciences and Technology, https://doi.org/10.1007/978-3-030-88644-8_5

To achieve a circular economy, it is necessary for the WM and the manufacturing sectors to work together. An integrated approach as a strategy for achieving a circular economy is needed in the WM and manufacturing sectors. In the case of the plastic industry, appropriately 4% of the annual petroleum manufactured from petrochemicals is directly converted into plastics. With 50% of the products manufactured from plastic materials being used in packaging products, this implies a continuous demand for raw materials. Incorporation of the IWS into formalized systems is a strategy that can be used to close the loop in the plastic industry and the WM sector. According to Agarwal et al. [3], most recovery and recycling of wastes remains an informal activity in developing economies and therefore, incorporation of the IWS into formalized systems can sustain the recovery and recycling activities. In 2005, Agarwal et al. conducted a study on MSW recycling and the connected markets in India. The researchers concluded that a hierarchy of recycling dealers consisting of waste recyclists performs an important role in SWM. Furthermore, the research indicated that recyclists needed to be integrated into WM formalized systems at local and urban frameworks. The study evaluated and proposed two models to the local municipal corporation for formalizing the informal recycling sector. In a study conducted by Devi and Satyanarayana [12], promoting and organizing microenterprises (IWS) is an effective strategy for extending affordable services to urban communities. Further, in a study conducted by Wilson et al. [55], training and organizing informal recyclers by converting them into microenterprises is an effective strategy for upgrading their capability of value addition to recovered materials and therefore contributing to SSWM.

The findings from the studies conducted by Wilson et al. [55], Devi and Satyanarayana [12] and Agarwal et al. [3] indicate that incorporation of the IWS into the formal microenterprises can contribute to SWM. Also, the studies provide evidence for incorporating the IWS into formalized systems. In addition, the studies have highlighted strategies for improving the lives of the IWS as a way of influencing more waste recoveries. More studies allude that integration of the informal sector into a formalized sector can contribute to SSWM. In a study that focused on the consumption, collection and recycling system for PET bottles in China [59], the results revealed that the informal sector recovers 90% of the PET bottles. The study also recommends the integration of the informal sector into formalized systems through the hiring of the IWCs. Haan et al. also indicated that converting the informal sector into microenterprises by training and organizing them is an effective strategy for upgrading value addition skills needed in the waste recycling industries. A study conducted by Mwanza et al. [35] outlined various strategies the IWS use to add value to recovered PSWs. All these studies indicate that incorporating and training the informal sector can contribute to SWM and value addition. In spite of the studies not focusing on the possibilities and limitations of implementing the recommendation of incorporating the IWS into formalized systems, all the studies reviewed were conducted in developing economies and this is of great significance. Further [60] alluded that improvement of the socioeconomic circumstances of the marginalized IWCs can be attained through formal–informal sector alliances and this has a direct positive effect on the services of WM in a city.

**Fig. 5.1**   Informal waste sector contribution to resource efficiency [19]

To achieve SWM for PSWs, incorporation of the IWS can contribute to the sustainment of the three pillars of sustainability. From an economic sustainability aspect, the IWS contributes to the reduction of economic costs that should be incurred by the formal waste sector (FWS). Ezeah et al. [14] indicated that the services provided by the IWS contribute to reducing the cost incurred by the FWS. The social sustainability aspect shows that a number of social benefits are gained through waste recovery by the IWS. According to Gunsilius et al. [19], income and work are generated from well-managed informal waste activities for many people in developing economies. Approximately 80% of the wastes recovered by the IWS in emerging and developing economies contribute to their livelihoods [14]. The wastes recovered from the environment are dangerous; therefore, the activities performed by the IWS sustain the environment and thus, the contribution of the IWS to the environmental sustainability pillar. Figure 5.1 depicts a schematic diagram of the contributions of the IWS to sustainable resource recovery.

Waste managers, engineers and entrepreneurs should understand that the activities performed by the IWS are important and contribute to sustaining the resources in a circular economy. However, these activities need to be performed properly by formalizing the IWS. Formalization of the IWS is a factor that can drive sustainable PSWs management. In a review study conducted by Mwanza et al. [34], the drivers for sustainable PSW management, include incorporation of the IWS into a formalized system. More studies have recommended the incorporation of the IWS into formalized systems ([15, 47]; Medina et al. [61]).

## 5.2  Economic Incentives

To achieve SSWM, a number of WM drivers are necessary and these should not work in isolation. With regards to other environmental actions, improvement of recycling waste activities is dependent on collective action while environmental outcomes are not exemptable [13, 26]. Economic incentives strategies have been applied in a number of countries to influence waste recovery activities. In a study performed by Yau [58] in Hong Kong, the results revealed that there exists a significant positive relationship between reward schemes and the weight of household recyclables collected while other variables are kept constant. Further, the study suggested that economic incentives promote WM activities in Hong Kong. Though the study was conducted in Hong Kong, the results are applicable in other developing economies with similar WM strategies. Attaching economic value to recovered wastes continues to contribute to high recovery rates [2, 10].

In developing economies, most local authorities in charge of WM encounter several challenges [9, 20, 30, 42]. In most cases, operating costs are a challenge because donors, environmental ministry, private company etc. provide external capital investment to the local authorities to operate landfills. Most local authorities in developing economies accept the operation of landfills without understanding the financial implications involved. For developed economies, landfilling is a priced ecosystem service and works as an economic incentive to the local authorities. The local authorities benefit from the funds paid by the waste generators and this amount is charged based on the actual weight of waste handed in for collection. The pricing of landfilling works as an economic incentive to the local authorities and also provides a platform for waste generators to segregate the wastes (PSWs).

Scheinberg [42] alluded that the pricing of landfills in developed economies has created economic incentives and enabled high performance among recycling entities. The more the waste generators dispose of, the more money is charged and this has created an opportunity for more materials destined for disposal to be recovered by the local authority.

In most developing economies, pricing of disposal is not based on the weighting system and this has created an increase in the quantity of waste disposal. The non-existence of a weighting system for waste destined for disposal has created a scenario in which an increase in the quantity of wastes is chasing the decreasing money in the local authority system. This has resulted in the formation of unmanaged dumpsites in developing economies [6, 19].

Waste managers and engineers should develop sustainable recovery systems that attract value in wastes. For example, in most developing economies, PSWs are recovered because they are valuable. According to Mutha and Pokharel [31], the implementation of economic benefits contributes to the take-back system of EoL products. In the recycling and recovery industry of PSWs, the value attached to the EoL products drives the supply chain network. In this regard, the supply chain network should be designed to optimize recovery rates. A study conducted by Mwanza [33] designed an African reverse logistic (RL) model for PSWs and all the key stakeholders in the supply chain were integrated. A number of other studies have designed RL models for EoL products [7, 16] and the reason behind the design is attributed to the value in the wastes. Therefore, the value in wastes is an economic driver and this should be used to influence recovery and recycling rates.

## 5.3  Appropriate Technology Considerations

PSWs have environmental effects and the majority of the developing economies are still struggling with sustainable approaches for managing them. In addition, recovery systems for PSWs do not quantify and characterize the wastes. For example, handling equipment (i.e. for sorting); vehicles for transportation; and recycling machinery do not match the PSWs characteristics.

The application of technology for recycling and recovering PSWs should be contextual. For example, most developed economies have advanced in technology application in the recycling industry [43] and this is not the case for developing economies. A study conducted by Mwanza and Mbohwa [32] revealed that most of the plastic recycling industries in Zambia use mechanical recycling technologies. Therefore, the systems for recycling should be upgraded and designed contextually. For example, a number of waste-handling vehicles underperform in developing economies. A study conducted by Klundert [52] alluded to the importance of understanding the contextual applicability of SW machinery. This is important for waste managers and engineers to implement in the SW projects.

It is interesting to understand that technological advancement has contributed to promising recycling rates in developed economies [43] and this is a call to developing economies to understand the applicability of advanced recycling technologies. Lessons from a study conducted by Plastic Recycling [28] showed that technological advancements in the plastic recycling industry are a sustainable driver. Several countries such as Lavia, The Netherlands, Germany and Japan affirmed that technological advancements in the recycling of plastics have contributed to high recycling rates. Hopewell et al. [21] also indicated that technological advancements are critical to recycling rates in the plastic industry.

From the fact that adequate funds are rarely available to implement effective use of technology in developing economies, managers and engineers should ensure that technologies sourced from developed economies fit into the context of application. Troschinetz et al. [49] noted that many developing economies are not adequately funded to enable them to acquire and implement appropriate technology. As a result, many developing economies resort to utilizing manual labor in place of technical machinery. Utilization of manual labor in place of technical machinery normally compromises the quality of the PSWs recovered and this reduces the targeted quantities. BIO-Intelligence [43] points out that innovations are important aspects in the recovery industries and many technologies are still required in the plastic industry. In essence, this is a call to developing economies to identify technologies that are applicable contextually. Already mechanical recycling is the most used plastic recycling technology [4, 35].

## 5.4  Societal Participation and Awareness in Recycling Schemes

Public awareness, participation and cooperation in developed economies are not comparable to developing economies. Consequently, most of the urban cities are generally clean in developed economies. In developing economies, implementation of cleanliness measures is not adequate and most communities bins are usually designated at fixed stations where residents are required to dispose of their wastes on a necessity basis [51]. In most cases, public participation is limited in developing economies. A study conducted in Indonesia by Zakianis et al. [62] revealed that public participation in developing economies is low. As a result of poor participation by residents, waste is normally littered around community bins and illegal dumpsites continue to mushroom at an alarming rate.

Several studies have focused on awareness and participation as drivers of waste recovery and recycling programs in communities [32, 54]. These studies concluded by affirming the relevance of awareness and participation in recovery and recycling programs. Nevertheless, it is important to provide adequate information to the community to enable them to participate in the recycling programs.

Public awareness of intended community waste recycling and recovery schemes is a key that can contribute to participation. According to Singhirunnusorn et al. [46] continuous provision of awareness and information on proper SWM and the environment is primary to the attainment of any community recovery and recycling project. In reality and practice, systems efficiency is proportional to citizens' participation in SWM systems.

In PSWs management activities, waste managers, engineers and other experts should design recovery and recycling systems that integrate societal participation and awareness as the drivers. For example, in Zambia, a campaign to keep Zambia clean was launched and the last day of each month is mandatorily set for the citizens to participate. Integration of all the citizens has contributed to participation rates. Awareness of the benefits of the campaign continues to drive the launch of the Keep Zambia Clean Campaign. A study conducted by Mwanza [33] on an African reverse model for PSWs tested the significance of knowledge and awareness on household's participation in recycling programs. Information dissemination on the benefits of PSWs recycling was rated 4.58 on a scale of 1–5. Further, in a study conducted by Omran et al. [37] in Indonesia, public awareness through radio, television and campaigns was recommended as a factor that can increase public participation.

Another avenue for waste managers to explore is awareness of PSWs recycling in learning institutions as the drive for a circular economy. According to the Brundt-land Report, sustainable development of communities meets the needs of the existing societies without compromising the needs of future societies. This implies the need to integrate learning institutions in sustainable recycling programs. For example, by 2050, the amount of waste generation is projected to increase to 27 billion [24]. Despite the projected increases in waste generation, active participation and aware-ness on the merits of PSWs d recycling among learning institutions can work as a driver.

In order to improve public participation in recovery and recycling projects, waste managers, engineers and municipalities should focus on the following:

- Engage and educate the community on household PSWs recycling businesses;
- Inform the public of the available support initiatives available to PSWs collectors;
- Conduct exhibitions to the public on waste minimization strategies;
- Promote the use of recycled products;
- Improve material recovery facilities (MRFs). This aspect has lagged behind in developing economies and requires active participation from the public. In most cases, MRFs are underdeveloped, non-existence or underutilized.

## 5.5 Regulations and Legislations

Regulations and legislations are the key drivers for SWM. In several developed economies, regulations and legislations have shaped the WM hierarchy. In Japan, the country with one of the highest recycling rates, extended producer responsi-bility (EPR) has been implemented [59]. EPR has affected recycling in many devel-oped economies like The Netherlands, The USA and Australia. Xevgenos et al. [57] conducted a study on success stories for MSW recycling and the study revealed that regulations and legislations contribute to sustainable recovery rates in developed economies.

Policy development in the WM arena is important and requires an integrated approach that covers aspects of cleanliness, preservation of environmental quality, maintenance of acceptable public health measures and sustained finance provisions. Policies should outline sustainability with respect to the availability of landfill space for future generations. To achieve this, financial provisions and technological inter-ventions should always accompany policy development in WM. In most European member states, the EPR principle mandates all economic operators selling packaging products on the market to be responsible for the management of the EoL packaging products [36]. Further, the development of the WM systems for packaging wastes is the responsibility of the operators. Normally the systems comply with the targeted recovery and recycling rates set by European law.

Implementation of regulations and legislations in most developing economies is faced with a number of difficulties. A study by Manaf et al. [27] on challenges and opportunities in WM highlights that inadequate legal provisions are one of the listed challenges faced by developing economies in the WM sector. For example, in the developing economy of Zambia, the government reinforced the EPR on plastic packaging products as a drive for SWM (ZEMA [63]). The re-enforcement of EPR in Zambia was directed toward the sustainable management of PSWs. It is important for waste managers, engineers and other stakeholders to understand that enforcement of EPR requires an integrated approach with all the stakeholders involved in the supply chain. Lessons from developed economies such as Japan should be noted by understanding how they have implemented EPR to achieve high recycling rates. According to Zhang and Wen [59], a reverse logistics framework was designed for the stakeholders involved in the recovery of PET bottles and other EoL post-consumer plastic products in Japan. The RL framework was designed with an integrated EPR regulation and this has worked in contributing to high recycling rates.

According to the World Bank, utilization of incentives for the purposes of waste recycling and reduction is a key to disposal costs savings. For example in the plastic industry, the introduction of packaging taxes, landfill bans and promotion of recycling in many developed and emerging economies has contributed to the positive results on recycling rates and waste reduction. It is necessary for waste managers, engineers and key stakeholders in the WM sector to understand that the government policies on WM and recycling enforcement and implementation vary from nation to nation (Troschlnitz [49]).

Government policies that refer to legislation and regulation implementation in developing economies lack effective enforcement and a number of studies have alluded to this [5, 39, 44]. As a result of this, the utilization of incentives to support government policies on WM and recycling activities has worked in some countries globally. For example, waste collection systems and MRFs such as recycling banks or garbage banks have worked in Thailand. According to the World Bank, community and school garbage banks have worked well in influencing recovery and recycling rates in Thailand. Incentives in the form of cash or reward points are given to the participants. In Japan, EPR with the integration of the IWS and the households continues to contribute to high recycling rates [59].

Regulations and legislations in the WM sector should be aligned to a legal framework that is in line with the national policy. This is an important factor as most legal documents are usually formulated as a means of assigning responsibility. The scope of the legal framework on WM should be broadened in order to trace accountability and responsibility. In the case of PSWs recovery and recycling, the targets and specified timeframes should be clearly defined in the legal frameworks. Further, the frameworks should facilitate the operation and planning of the recycling system. For example, the definition of the term "PSWs" should be clearly outlined and provide the critical drivers useful to address high recovery and recycling rates. An RL model designed for PSWs by Mwanza [33] optimizes the projected PSWs to be recovered for and recycling among the integrated stakeholders by utilizing the drivers.

## 5.6 Collection and Segregation Systems

Waste collection is an important aspect of the WM system. However, in the era of sustainability and the circular economy, it cannot be discussed in isolation of waste segregation. Numerous studies have assessed the impact of SW collection systems on sustainability and the findings from these studies call for waste managers, engineers and stakeholders to understand waste collection systems.

Dahle and Lagekvist [11] allude that waste collection systems are classified into property-close and drop-off systems. Property-close waste collection systems are divided into kerbside and door-to-door collection systems. For households that are serviced with the kerbside collection system, containers and instructions are provided. In the case of door-to-door waste collection systems, households are given waste containers and advised to keep their containers at their premises. The difference between drop-off collection systems and property-close is that in drop-off systems, residents deliver the recyclables and wastes to the drop-off centers and points [29].

Buy-back centers are another type of waste collection system in which households deliver recyclables and financial incentives are provided depending on the amount delivered. Deposit–refund systems contribute to the recovery and return of reusable and recyclable products. For this system, an amount of money is kept by the seller of the packaged product until the buyer returns the EoL container. Mostly, the deposit–refund system works well in the beverage industry.

Numerous studies have focused attention on waste collection systems [11, 17, 18, 23, 25, 41, 48]. These studies have assessed waste collection systems under the social, economic and environmental parameters of sustainability. Despite the differences in the approaches of the studies, waste managers and engineers should understand that waste collection systems are a critical parameter for achieving sustainable management of PSWs. It is important that waste managers and engineers consider socioeconomic factors before the implementation of a particular type of waste collection system. Further, waste collection should be adjoined to waste segregation. Household engagement in waste segregation activities continues to contribute to the sustainable recovery of PSWs in most developing economies [59].

## 5.7 Household Education

A study conducted by Afroz et al. [1] pointed out that educational campaigns are one of the key factors for boosting waste diversion. In Brazil educational campaigns have contributed to the massive waste recovery. It is important for households to be educated on WM issues. Knowledge on WM provides linkages between health, environment, sanitation, waste handling and households [49]. In countries such as Thailand, Mongolia and Sri Lanka, household education on SWM is perceived as an incentive for influencing participation in waste recovery and minimization activities [8].

In Thailand, projects by non-government and government organizations, such as "Thinking about and Saving Resources" and "Waste Minimization," focus on increasing household education on waste reduction, recycling and reusing. In developing economies, municipalities and private waste companies should use household education on WM as a key driver to SWM. A community that is educated on the benefits of WM is likely to participate in WM projects. Tunmise and Seng [50] pointed out that household support to waste recovery and recycling programs is essential. Active and sustained participation of the households results in successful programs [22].

## 5.8  Institutional Arrangements

Institutions refer to the legal rules, norms and conventions of a society that provide meaningful stability for coordination that is essential to support values and protect interests [53]. Institutional arrangements are important for achieving sustainable WM. For example, the municipal corporation is responsible for the sustainable management of SW and this has resulted in the involvement of the local government operations. The government has a responsibility as far as financial and institutional support are concerned. However, in developing economies, the majority of the studies have revealed that the local governments responsible for managing SW are inadequately funded. As a strategy for improving operational efficiencies in the local government, the private sector has been fully integrated into the system [45].

Institutional arrangements in PSWs management should be among the key stakeholders, which include the local government, the private sector and the community, and these can be further categorized into legal, financing and regulatory components [40]. Figure 5.2 depicts the scenario of institutional arrangement for achieving sustainable WM. The key stakeholders presented in the study by Rico et al. [40] are appropriate and relevant. However, in developing economies where the majority of PSWs recovery is conducted by the IWS, integration of the IWS should be driven and encouraged by waste managers and engineers. The integration of the informal waste sector into a formalized system is supported by numerous studies, as discussed in Sect. 5.1.

## 5.9  Training the IWS

The FWS and IWS form the two pillars that support the urban economy. Even though there are no equal economic contributions from the two sectors, the IWS makes significant contributions in WM issues. Zia and Devadas [60] affirmed that the IWS not only strengthens the economy of an urban society, it also provides an invaluable service as well as a firm standing. Despite their contributions, many of the IWCs involved in SWM practices such as recycling suffer a number of challenges. Wilson

**Fig. 5.2** Major stakeholders in SSWM [40]

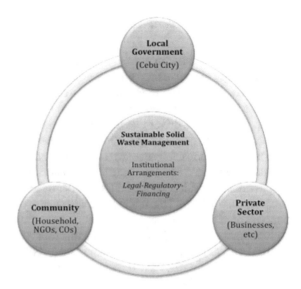

et al. [56] highlighted that a number of public policies in WM have been driven by traditional norms of controlling environmental and public health consequences of poor WM and this has shown the legal context of the informal recycling operations. The study highlighted that public policies governing the operations of the IWS are largely negative in many countries. Even though issues of public health are key drivers in WM, the majority of the stakeholders (i.e. waste pickers) directly involved in WM suffer from health-related illnesses. The question to be asked is how health-related issues in informal waste recycling should be overcome? Several studies have recommended training of the IWS on aspects of waste sorting. Agarwal et al. [3] recommended training of the IWS on personal hygiene as the majority of the recyclists are unaware of the repercussions of sorting garbage without adhering to proper safety guidelines. Further, Ezeah et al. [14] highlighted the need for comprehensive training programs on personal hygiene for all the IWCs involved in waste reuse and recycling activities. The establishment of training programs for the IWS can promote more waste recovery as those involved will have the drive to do so. From the fact that recycling forms the largest element of the IWS in WM activities [60], there is a need to use training programs as driving mechanisms for SSWM. If the informal sector is trained on waste sorting and on hygiene factors, this can motivate other stakeholders to join in waste recovery. It is thus imperative for potential improvement measures to be added to the existing recycling sectors in order to enhance and promote the conservation of resources and energy for future generations.

### 5.9.1 Local Recycled Material Market

The availability of local markets for recyclable wastes is a key driver to sustainable waste recovery and recycling. For example, Jamaica with 40% of recyclable wastes lacks viable markets due to low quantities of recoverable wastes [38]. Profitability and existence of markets systems that depend on recycled raw materials throughout the supply chain is a driver that influences recovery in developing economies. The truth is that the majority of waste recovery is performed by the IWS in developing economies, and the establishment of sustainable buy-back centers can boost recovery. In a study conducted by Mwanza [33], the majority of PSWs recovered were sold to the intermediate dealers by the IWCs. The study also reveals that most of the IWCs are dumpsites pickers. This implies that the majority of PSWs are not recovered but disposed of at the dumpsites.

To promote sustainable recycling systems for PSWs, some developed economies (i.e. the USA, Latvia and Denmark) export their PSWs to China [28]. This aspect has continued to promote the PSWs recovery in these nations. For developing economies, the establishment of integrated supply chain networks for recycled PSWs is required. Already, a number of IWCs are actively involved in the recovery process but markets for recycled materials lag behind. Mwanza [33] reveals that the existence of local markets determines the recyclability of PSWs from the recovery industry's perspective. BIO-Intelligence [43] points out that recycling and recovery rates for PSWs have improved in developed economies as a result of linkages amongst plastic recyclers, manufacturers and other players in the supply chain. Local market establishment is a key driver to sustainable PSWs recycling and recovery. For example, 15 million tons of Chinese original PSWs were recycled in 2011 as a result of the existence of domestic markets.

Ezeah et al. [14] reveal that reusability is increasing the value of wastes. For example, PET bottles recovered by the IWCs are reused for purposes such as storage of fluids other than recycling purposes alone. With the increase in the application of plastic materials, a number of products are being manufactured and this has a direct impact on the number of recyclables generated. From the fact that the recovery process is normally left in the hands of the informal sector in developing economies, the majority of the PSWs contribute to environmental degradation. In Iran, the revenue of 44 million dollars would have been earned if an official recycling system for cardboards and paper was established. This example applies to PSWs considering the different purposes for which it is used.

## References

1. R. Afroz, A. Rahman, M.M. Masud, R. Akhtar, The knowledge, awareness, attitude and motivational analysis of plastic waste and household perspective in Malaysia. Environ. Sci. Pollut. Res. **24**, 2304–2315 (2017)

2. P. Agamathu, K.M. Khidzir, F.S. Humid, Drivers of sustainable waste management in Asia. Waste Manage. Res. **27**, 625–633 (2009)
3. A. Agarwal, A. Singhmar, M. Kulshrestha, A.K. Mittal, Municipal solid waste recycling and associated markets in Delhi, India. Resour. Conserv. Recycl. **44**, 73–90 (2005)
4. S.A. Al-Salem, P. Lettieri, J. Baeyens, Recycling and recovery routes of plastic solid waste (PSW). a review. Waste Manage. **29**, 2625–2643 (2009)
5. I. Arbulú, J. Lozano, J. Rey-Maquieira, The challenges of municipal solid waste management systems provided by public-private partnerships in mature tourist destinations: the case of Mallorca. Waste Manage. **51**, 252–258 (2016)
6. J.M. Ball, L. Bredenhann, *Minimum Requirements for Waste Disposal by Landfill*, 2nd edn. (Department of Water Affairs for Republic South Africa, 1998)
7. X. Bing, J.M. Bloemhof-Ruwaard, J.G.A. van der Vorst, Sustainable reverse logistics network design for household plastic waste. Flex. Serv. Manuf. J. **26**, 119–142 (2014)
8. O. Buenrostro, G. Bocco, Solid waste management in municipalities in Mexico: goals and perspectives. Resour. Conserv. Recycl. **39**, 251–262 (2003)
9. T.M. Coelho, PET containers in Brazil: opportunities and challenges of a logistics model for post-consumer waste recycling. Resour. Conserv. Recycl. **3**(55), 291–299 (2011)
10. F. Contreras, S. Ishil, T. Aramaki, Drivers in current and future municipal solid waste management systems: cases in Yokohama and Boston. Waste Manage. Res. **28**, 76–93 (2010)
11. L. Dahlén, A. Lagerkvist, Evaluation of recycling programmes in household waste collection systems. Waste Manage. Res. **28**, 577–586 (2010)
12. K. Devi, V. Satyanarayana, Financial resources and private sector participation in SWM in India. Indo-US Financial Reform and Expansion (FIRE) Project (New Dehli, 2001)
13. J.W. Everett, J.J. Pierce, Curbside recycling in the U.S.A.: convenience and mandatory participation. Wasre Manag. Res. **11**, 49–61 (1993)
14. C. Ezeah, J.A. Fazakerley, C.L. Roberts, Emerging trends in informal sector recycling in developing and transition countries. Waste Manage. **33**, 2509–2519 (2013)
15. F. Fei, L. Qua, Z. Wena, Y. Xueb, H. Zhang, How to integrate the informal recycling system into municipal solid waste management in developing countries: based on a China's case in Suzhou urban area. Resour. Conserv. Recycl. **110**, 74–86 (2016)
16. G.L. Ferri, G.L.D. Chaves, G.M. Ribeiro, Reverse logistics network for municipal solid waste management: the inclusion of waste pickers as a Brazilian legal requirement. Waste Manage. **40**, 173–191 (2015)
17. A. Gallardo, M.D. Bovea, F.J. Colomer, M. Prades, M. Carlos, Comparison of different collection systems for sorted household waste in Spain. Waste Manag. **30**, 2430–2439 (2010)
18. A. Gallardo, M.D. Bovea, F.J. Colomer, M. Prades, Analysis of collection systems for sorted household waste in Spain. Waste Manag. **32**, 1623–1633 (2012)
19. E. Gunsilius et al., *Recovering Resources, Creating Opportunities: Integrating the Informal Sector into Solid Waste Management* (Aksoy Print, Eppelheim, Germany, 2011)
20. J. Gutberlet, *Recovering Resources—Recycling Citizenship: Urban Poverty Reduction in Latin America* (Ashgate, Aldershot, 2008)
21. J. Hopewell, R. Dvorak, E. Kosior, Plastics recycling: challenges and opportunities. *Phil* (2009)
22. A. Ittiravivongs, Household waste recycling behavior in Thailand: the role of responsibility, in *International Conference on Future Environment and Energy. International Proceedings of Chemical Biological and Environmental Engineering* (2012, 21–26)
23. A.X. Karagiannidis, G. Perkoulidis, N. Moussiopoulos, Assessing the collection of urban solid wastes: a step towards municipality benchmarking. Water Air Soil Pollut. Focus **4**, 397–409 (2004)
24. T. Karak, R.M. Bhagat, P. Bhattacharyya, Municipal solid waste generation, composition, and management: the world scenario. Crit. Rev. Environ. Sci. Technol. **42**(15), 1509–1630 (2012)
25. A.W. Larsen, H. Merrild, J. Moller, T.H. Christensen, Waste collection systems for recyclables: an environmental and economic assessment for the municipality of Aarhus. Waste Manag. **30**, 744–754 (2010)

26. M. Lubell, A. Vedlitz, S. Zaharan, L. Alston, Collective action, environmental activism, and air quality policy. Polit. Res. Q. **59**(1), 149–160 (2006)
27. A.L. Manaf, M.A. Samah, I.M.N. Zukki, Municipal solid waste management in Malaysia: practices and challenges. Waste Manage. **29**, 2902–2906 (2009)
28. Managment, Plastic Waste. *An Introduction to Plastic Recycling* (Institute of Plastic Waste Management, 2009)
29. C. Mbande, Appropriate approach in measuring waste generation, composition and density in developing areas. J. S. Afr. Inst. Civ. Eng. **45**(3), 2–10 (2003)
30. M. Medina *The World's Scavengers: Salvaging for Sustainable Consumption and Production* (AltaMira Press, 2007)
31. A. Mutha, S. Pokharel, Strategic network design for reverse logistics and remanufacturing using new and old product modules. Comput. Ind. Eng. **56**(1), 334–346 (2009)
32. B.G. Mwanza, C. Mbohwa, *Technology and Plastic Recycling: Where Are We in Zambia, Africa?" IAMOT 2019—Managing Technology for Sustainable and Inclusive Growth Conference Proceedings* (Excel India Publishers, Mumbai, 2019), pp. 964–971
33. B.G. Mwanza, An African reverse logistics for plastic solid wastes. University of Johannesburg, Ph.D. Dessertation, Johannesburg, 2018
34. B.G. Mwanza, C. Mbohwa, Drivers to sustainable plastic solid waste recycling: a review. Procedia Manuf. **8**, 649–656 (2017)
35. B.G. Mwanza, C. Mbohwa, A. Telukdarie, C. Medoh, Value addition to plastic solid wastes: an informal waste collectors' perspective. Procedia Manuf. **33**, 391–397 (2019)
36. OECD, *Extended Producer Responsibility: A Guidance Manual for Governments* (OECD, Paris), 2001
37. A. Omran, A. Mahmood, H. Abdul Aziz, G.M. Robinson, Investigating households attitude towards recycling of solid waste in Malaysia: a case study. Int. J. Environ. Res. **3**(2), 275–288 (2009)
38. P.S. Pendley, *Feasibility and Action Plan for Composting Operation Incorporating Appropriate Technology at Riverton Disposal Site, Kingston, Jamaica* (Master of Science in Environmental Engineering, Kingston) (2005)
39. D. Reinhart, C.S. Bollard, N. Berge, Grand challenges—management of municipal solid waste. Waste Manage. **49**, 1–2 (2016)
40. C.A. Rico, D.A. Nestor, M.R. Carmelita, Institutional arrangements for solid waste management in Cebu City, Philippines. J. Environ. Sci. Manag. **15**(2), 74–82 (2012)
41. S. Rodrigues, G. Martinho, A. Pires, Waste collection systems. Part A: a taxonomy. J. Clean. Prod. **113**, 374–387 (2016)
42. A. Scheinberg, Value added, modes of sustainable recycling in the modernisation of waste management systems. Ph.D. Dissertation Wageningen University, The Netherlands. Gouda. The Netherlands: WASTE, 2011
43. Service BIO-Intelligence, Study on an increased mechanical recycling target for plastics, in *Recyclers Europe Recyclers Europe for Final Report Prepared for Plastic* (2013)
44. R.H. Sharma, B. Destaw, T. Negash, L. Negussie, Y. Endris, G. Meserte, B. Fentaw, A. Ibrahi, Municipal solid waste management in Dessie City, Ethiopia. Manage. Environ. Qual. Int. J. **24**(2), 154–164 (2012)
45. V.A. Shekdar, Sustainable solid waste management: an integrated approach for Asian countries. Waste Manage. **29**, 1438–1448 (2009)
46. W. Singhirunnusorn, K. Donlakorn, W. Kaewhanin, Contextual factors influencing household recycling behaviors: a case of waste bank project in Mahasarakham municipality. Procedia. Soc. Behav. Sci. **36**, 688–697 (2012)
47. D. Storey, L. Santucci, R. Fraser, J. Aleluia, L. Chomchuen, Designing effective partnerships for waste-to-resource initiatives: lessons learned from developing countries. Waste Manage. Res. **33**(12), 1066–1075 (2015)
48. C.A. Teixeira, C. Avelino, F. Ferreira, I. Bentes, Statistical analysis in MSW collection performance assessment. Waste Manag. **34**, 1584–1594 (2014)

49. A.M. Troschinetz, J.R. Mihelcic, Sustainable recycling of municipal solid waste in developing countries. Waste Manage. **29**(2), 915–923 (2009)
50. A.O. Tunmise, L. Seng, Municipal solid waste management: household waste segregation in Kuching South City, Sarawak, Malaysia. Am. J. Eng. Res. (AJER) **3**(6), 82–91 (2014)
51. UNEP, State of the environment (2001), http://www.eapap.unep.org/reports/soe/. Accessed 2016
52. A. van de Klundert, Integrated Sustainable Waste Management: the selection of appropriate technologies and the design of sustainable systems is not (only) a technical issue, in *CEDARE/IETC Inter-Regional Workshop on Technologies for Sustainable Waste Management* (Alexandria, Egypt, 1999), 1–16
53. A. Vatn, Rationality, institutions and environmental policy. Ecol. Econ. **55**(2), 203–217 (2005)
54. P. Vicente, E. Reis, Factors influencing households' participation in recycling. Waste Manage. Res. **26**, 140–146 (2008)
55. D.C. Wilson, C. Velis, C. Cheeseman, Role of informal sector recycling in waste management in the developing countries. Habitat. Int. **30**, 787–808 (2006)
56. D. Wilson, A. Whiteman, A. Tormin, *Strategic Planning Guide for Municipal Solid Waste Management* (World Bank, Washington, DC, 2001)
57. D. Xevgenos, C. Papadaskalopoulou, V. Panaretou, K. Moustakas, D. Malam, Success stories for recycling of MSW at municipal level. Waste Biomass Valor **6**, 657–684 (2015)
58. Y. Yau, Domestic waste recycling, collective action and economic incentive: the case of Hong Kong. Waste Manage. **30**, 2440–2447 (2010)
59. H. Zhang, Z.G. Wen, The consumption and recycling collection system of PET bottles a case study of Beijing, China. Waste Manage. **34**, 987–998 (2014)
60. H. Zia, V. Devadas, S. Shukla, Assessing informal waste recycling in Kanpur City, India. Manag. Environ. Qual. **19**, 597–612 (2008)
61. M. Medina, Globalization, development, and municipal solid waste management in third world cities. Available at http://depot.gdnet.org/cms/conference/papers/5th_pl5.2_martin_medina_martinez_paper.pdf Accessed 12 August 2016 (2002)
62. S. Zakianis, I.M. Djaja, The importance of waste management knowledge to encourage household waste-sorting behaviour in Indonesia. Int. J. Waste Resources **7**(04)
63. Zambia Environmental Management Agency, *Statutory Instrument No. 65 on Extended Producer Responsibility Regulations* (Duke Nicholas Institute for Environmental Policy solutions, LUSAKA, 2018)

# Chapter 6
# Policy Makers Responsibilities

## 6.1 Sustainability Pillars as Key Drivers in Waste Management

PSWs are a problem that requires the involvement of policy makers. The policy makers alone cannot change the current picture of PSWs management and therefore the government, corporate and non-governmental organizations (NGOs) engagements should provide a platform for tabling the economic, environmental and social drivers of managing PSWs. The government efforts should be directed toward discussing the international legally binding treaty on PSWs management. Further, the government should enforce the dissemination of information and strategies for effectively managing PSWs. At the corporate level, there are several initiatives for obtaining multinational corporations that plastic users and producers can work together to standardize the use of plastic material, separation and recycling. NGO's job is to push the government and companies to enforce the initiatives for managing PSWs sustainably. Another approach that NGOs can push is citizens' awareness of PSWs management. For example, a study conducted on knowledge and awareness on PSWs management [2] revealed the need to educate citizens on the sustainable management of PSWs.

Economic drivers have influenced the management of PSWs. In a study carried out by Mwanza ct al. [1] on the drivers for sustainable management of PSWs, incentives and value in waste are the drivers. For example, the majority of the IWCs are driven to recover wastes of their value. They recover to sell to recycling and manufacturing companies. This shows there is monetary value in recovering PSWs. The incentives provided to communities in the form of money and goods for the returned PSWs have influenced participation. For example, households will only recover PSWs because of the anticipated incentive that is provided to them. Manufacturing and recycling companies are driven to buy from waste collectors because of the reduced costs

B. G. Mwanza and C. Mbohwa, *Sustainable Technologies and Drivers for Managing Plastic Solid Waste in Developing Economies*, SpringerBriefs in Applied Sciences and Technology, https://doi.org/10.1007/978-3-030-88644-8_6

of buying PSWs compared to virgin plastic materials. Landfill disposal fees have influenced sustainable management of PSWs, for example in the EU nations, the fee for disposing waste at the landfill is comparable to recycling costs. This aspect has encouraged recyclers to participate in plastic recovery and recycling activities. Developed economies have changed the fixed charge on waste collected for disposal, instead, waste generators are charged based on the quantity generated. This option has seen waste generators being mindful of how much wastes they generate. The above economic drivers contribute to the sustainable management of PSWs in developing and developed economies.

Environmental drivers continue to influence stakeholder participation in WM initiatives. Zaman [3] alludes that environmental awareness, worldwide climate change and movement are the environmental drivers for sustainable Mwanza et al. [1] indicated that global warming, climate change and environmental protection are environmental drivers for SSWM.

The social drivers for sustainable management of solid wastes are cardinal for stakeholders' awareness. Zaman [3] concludes that personal behavior, consumption and generation, and local WM practices are the key social drivers. Figure 6.1 depicts the economic, environmental and social WM development drivers.

Zaman [3] shows that several drivers categorized as economic, social and environmental are cardinal to the development of SSWM systems. In addition, the figure shows that producer responsibility and consumer accountability drivers form the subset for the three aspects of sustainability. This implies the need for producers to

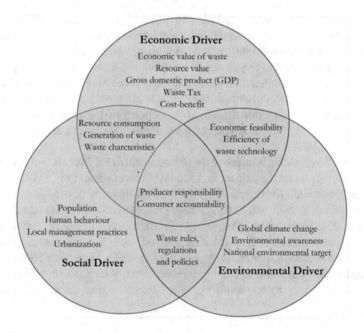

**Fig. 6.1**  Waste management development drivers [3]

be responsible for the EoL products. This means that the manufacturers and distributors of producers should be responsible for the processes that occur after the product's life has ended. In relation to the consumers, accountability should be taken in the way EoL products are managed. These drivers were revealed in the context of a developed economy. However, waste managers, engineers and other parties interested in WM systems can apply these drivers in the context of a developing economy. Engagement of the key stakeholders is cardinal for the successful implementation of strategies and systems. For instance, Japan has implemented a sustainable system for managing PSWs [4]. The system has integrated the key stakeholders as well as the drivers that influence the successful management of waste in Japan's context.

## 6.2 Integrating the PSWs Drivers and Technologies for Sustainable Management of PSWs

Based on the evaluation of the studies in the previous chapters, sustainable management of PSWs can only be achieved through an integrated system. Several studies have revealed a number of drivers required to push the management of PSWs to sustainability. Therefore, focusing on the drivers alone without paying attention to the technologies presents another gap to achieving sustainability. Figure 6.2 depicts the integration of the drivers and technologies for sustainable PSW management. The diagram shows that the drivers work at household, community and institutional levels.

The proposed integrated PSWs management system can help engineers, managers and policy makers to understand the relation of sustainability and the circular economy. The drivers and technologies continue to help in managing PSWs from the household, community and institutional levels. For companies and institutions, drivers such as economic incentives, technology considerations, collection and segregation, institutional arrangement, household education, regulations and legislations, and IWS acceptance can help boost PSWs recovery and management. These drivers can be analyzed within the context of each company or organization to determine if they can work.

In developing economies, several technologies are used for managing PSWs. These technologies have presented success stories that have driven and encouraged companies and communities to participate in PSWs management. Therefore, it is important for engineers, managers and policy makers to integrate the technologies and the drivers in order for the households and communities to understand from the sustainability pillars perspective. Further, the integrated PSWs management system can help managers identify the most fragile drivers of sustainability. In this case, integration of the IWS into formalized systems can influence the management of PSWs at community, household and institutional levels. This social aspect is important in developing economies because the majority of waste is recovered by the IWS.

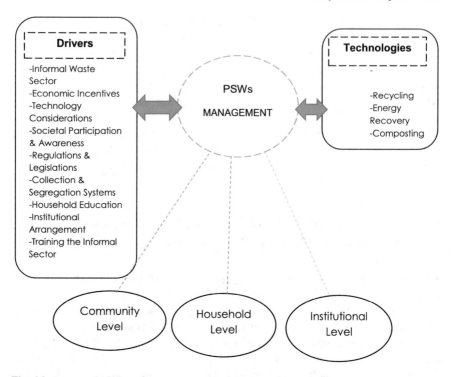

**Fig. 6.2** Integrated PSWs management systems (Author 2019)

# References

1. B. Mwanza, C. Mbohwa, Drivers to sustainable plastic solid waste recycling: a review. Procedia Manuf. **8**, 649–656 (2017)
2. B. Mwanza, C. Mbohwa, Knowledge and awareness on plastic solid waste management in Zambia: where are we? West Afr. Built Environ. Conf. Proc. WABER, Accra
3. A. Zaman, Identification of waste management development drivers and potential emerging waste treatment technologies. Int. J. Environ. Sci. Technol. **10**, 455–464 (2013)
4. H.A. Zhang, The consumption and recycling collection system of pet bottles: a case study of Beijing, China. Waste Manage. **34**, 987–998 (2014)

# Chapter 7
# Findings and Contributions to Sustainable Plastic Solid Waste Management

Sustainable management of PSWs is a challenge faced by developing economies. To this end, the book focused on various aspects that contribute to the sustainable management of PSWs in developing economies while using the success stories of developed economies.

An understanding of SSWM from a developing economy perspective is revealed in Chap. 1 and the different definitions of SW are evaluated. Redefining SW contributes to the body of knowledge in the waste management arena as it enables the relevant stakeholders to understand what is already known about SW from a different perspective. Insights to sustainable ways of managing PSWs are presented through the adoption of an SSWM model.

Sustainable and successful management of PSWs is inevitable without an understanding of the various kinds of PSWs. Chapter 1 contributes to the existing knowledge by presenting different challenges being faced in the management of PSWs in developing economies. In order to close the gap on the challenges facing developing economies to manage PSWs, applicable solutions adapted from success stories of developed economies are presented. Further, a comparison of the projected PSWs generation from an income perspective is discussed to show the urgency required to implement sustainable solutions.

Various technologies for managing SW have been developed and the majority of the advanced technologies have been implemented in developed economies. In this regard, current technological trends in the management of PSWs from a developing economy contribute to the existing theories. In a number of economies, mechanical recycling technology is the most utilized technology and this is an indication to the relevant stakeholders on the need to develop this technology in developing economies. Understanding of the technologies for managing PSWs is enhanced with discussions on the social, economic and environmental implications.

B. G. Mwanza and C. Mbohwa, *Sustainable Technologies and Drivers for Managing Plastic Solid Waste in Developing Economies*, SpringerBriefs in Applied Sciences and Technology, https://doi.org/10.1007/978-3-030-88644-8_7

Assessment of the indicators of sustainability is cardinal for ensuring the successful implementation of systems. Utilization of indicators provides a platform for identifying potential problems, development of workable policies and simplification of the systems. Indicators of sustainability are important to discuss because of the cause and effect relationship that exists among the complex environmental, economic and social aspects. Contributions in sustainable engineering have been presented by identifying the indicators that waste managers, engineers and other stakeholders should consider.

A number of factors influence the successful implementation of sustainable PSWs management systems. These factors exist in developed as well as developing economies. Chapter 5 contributes to the existing body of knowledge by highlighting the factors that influence the successful implementation of PSWs management systems. These factors are considered drivers because of the positive impact on the implementation of PSWs systems in developed economies. Further, the applicability of these drivers is emphasized by consideration of feasibility studies.

From an economic, environmental and social perspective, the drivers for sustainable management of PSWs are revealed in Chap. 6. The chapter contributes to the existing literature by highlighting the relevance policy makers hold in directing the sustainable management of PSWs. Further, an integrated PSWs management system is proposed. The proposed system contributes to the existing systems and theories for managing SWs and PSWs in particular.

Understanding sustainable technologies for managing PSWs from a developing economy perspective is necessary. Therefore, the book has contributed to the existing literature by taking into consideration various aspects that have worked in developed economies for application in development. The proposed application is not being considered in isolation of feasibility studies. Further, the area of sustainable PSWs requires more attention because of the prospects and barriers it presents.

One has to be imaginative and create 20–30 or more pages. This must be like all the previous chapters—critical work in your thesis and additional material collected but not used during your doctoral studies. It must also indicate suggested future research work.

Printed in the United States
by Baker & Taylor Publisher Services